Laboratory Manual
to accompany

WORLD OF
CHEMISTRY

SECOND EDITION
EXTENDED VERSION

John R. Blackburn
Georgetown College

Melvin D. Joesten
Vanderbilt University

Paul B. Langford
David Lipscomb College

John C. Craig
Consultant; West Point, Mississippi

Saunders Golden Sunburst Series

Saunders College Publishing
Harcourt Brace College Publishers

Fort Worth Philadelphia San Diego New York Orlando Austin
San Antonio Toronto Montreal London Sydney Tokyo

Joesten, Blackburn, Langford & Craig; Laboratory Manual to
accompany World of Chemistry, 2e.

ISBN 0-03-007442-8

567 095 7654321

Preface

The chemical sciences, like all areas of science, have undergone a tremendous information and technology explosion in recent years, and are continuing to experience exponential growth in these areas. Chemistry is a visual, dynamic science where experiential learning is as vital as theoretical pedagogy; and is a science where hands-on activities are not only desirable but are inseparable from lecture and textbook studies. The academic laboratory provides a wealth of opportunities for students to see and experience the reactions and concepts discussed in lecture. No chemistry class is complete without accompanying laboratory experiences. The experiments contained in this laboratory manual have been designed to help remove the aura of mystery surrounding the wonderful transformations which matter undergoes, and to illustrate the topics covered in an introductory chemistry course in order to make the ideas more concrete and real.

This new edition of a lab manual which in previous form has already served many students in their study of chemical principles has been revised with a number of goals directing the revision. As more attention is paid daily to environmental and safety concerns, we authors have tried to insure that the experiments and activities be as safe and "environmentally friendly" as possible. However, professors and students alike must recognize that it is impossible to insulate oneself from all materials that might be considered "hazardous" or of environmental concern. Indeed, many of the substances which are heavily regulated by safety agencies when they are used in the laboratory are in fact commonly used around the home, and are used and discarded without further thought by the average homeowner. Drain cleaners, cleaning solvents and paints, insecticides, and fuels are only a few of these products. It would be impossible to structure a semester's laboratory experiences for a student that do not incorporate the use of materials that could be hazardous if misused. We have added to this manual two new Labtext sections to help the student understand and deal with hazardous materials with their associated properties. Studying these sections should be an important part of the student's introduction to the laboratory in the first few weeks.

The experiments contained in this manual span the common range of types of laboratory experiences encountered in the daily life of most chemists. There are classical "wet" quantitative analyses, instrumental analyses, qualitative analysis, both organic and inorganic syntheses (making new substances from other ones), and a variety of experiments to show how important measurements on matter are made (molecular weights, densities, solubilities, physical properties, etc).

The experiments in this manual are experiments, not demonstrations. The goal for each experiment is that the student solve a problem by using his or her own observations and deductions. In some cases, students are encouraged to work together to stimulate discussion and share ideas in order to more fully understand the principles being illustrated by the experiment. In all cases, it is the intent of the authors that the students *have fun while learning*, and learn in a safe environment. The intent of this manual is not to train professional chemists, but to illustrate chemical principles to liberal arts students in order that they better understand the underlying role of chemistry in daily life.

No claim of originality is made for any of the experiments contained herein, although most have been either developed or adapted by the authors for their specific courses. Students at Vanderbilt University, David Lipscomb University, and Georgetown College have played an important role in the evolution of these experiments, which have been tested and improved through the years as a result of the efforts of these students.

Some experiments that had become "dated" or that were deemed by users of the manual to not be as useful as others have been deleted from this edition, and new experiments that incorporate more "everyday" chemical activities have been added. Many classical experiments that have proven good through the years have been retained, but all have been carefully reviewed and revised as needed. New features to all experiments include the addition of both pre-and post-lab questions for each experiment. The pre-lab questions will help the student prepare for the laboratory experiences, and the post-lab questions will help the students determine if they have understood the purpose and the results of their activities.

The authors wish to thank those authors and contributors who helped with the previous editions of this manual, but who chose not to be involved with this latest revision, and who are acknowledged in a different place in this work. Also, the aid of the many students and professors around the country who have used the manual and who have provided comments and insights on the various experiments is also gratefully acknowledged. It is through this process that texts improve and become ever more useful for the public interested in investigating any subject area, and it is our fervent hope that the users of this current edition continue to volunteer their suggestions for changes and improvements.

The Authors

ACKNOWLEDGEMENTS

The authors wish to acknowledge the following who contributed experiments that appeared in the first edition of the lab manual, several of which are found in this second edition: Mark M. Jones, David O. Johnston, John T. Netterville, James L. Wood, George E. Walden, John W. Dawson.

Contents

Introduction to the Laboratory

LABORATORY APPARATUS

Each student will have the use of a set of apparatus at an assigned work place. It should be remembered that each set of apparatus is used by other students in other laboratory sections.

1. Your work area and apparatus must be kept clean. Do not leave dirty equipment in your desk.

2. At the end of the laboratory period, you should clean your laboratory area with a damp paper towel.

3. Do not borrow apparatus from other desks. If you need extra equipment or an item that is not available at your assigned desk, obtain this from the stockroom.

4. When a piece of equipment is broken, go to the stockroom for replacement.

5. Large pieces of equipment (ring stands) are stored near your desk and some of the seldom used items will be set out on the side shelf.

6. Many of the typical kinds of laboratory apparatus are shown and named in Figure Intro–1.

LABORATORY SAFETY

The chemistry laboratory, like the automobile, can be operated in a relatively safe manner or can pose considerable risk if used improperly. The basic goal of a laboratory safety program is to reduce the risk of laboratory work to that involved in ordinary living. This is an achievable goal only with a good plan and excellent cooperation between instructors and students.

Three items of knowledge and understanding are of paramount importance as you approach every laboratory experience.

1. Know the location of each piece of safety equipment and understand its proper use.

2. Know in advance the points of danger in the planned laboratory procedure and understand both how to operate the plan safely and what to do in the worst-possible-case series of accident events.

3. Know that you do not know enough to do impromptu experimentation, or work without supervision, in a chemistry laboratory, and understand that even the best researchers should both work a planned procedure and have back-up help available from other professionals in case of unexpected trouble.

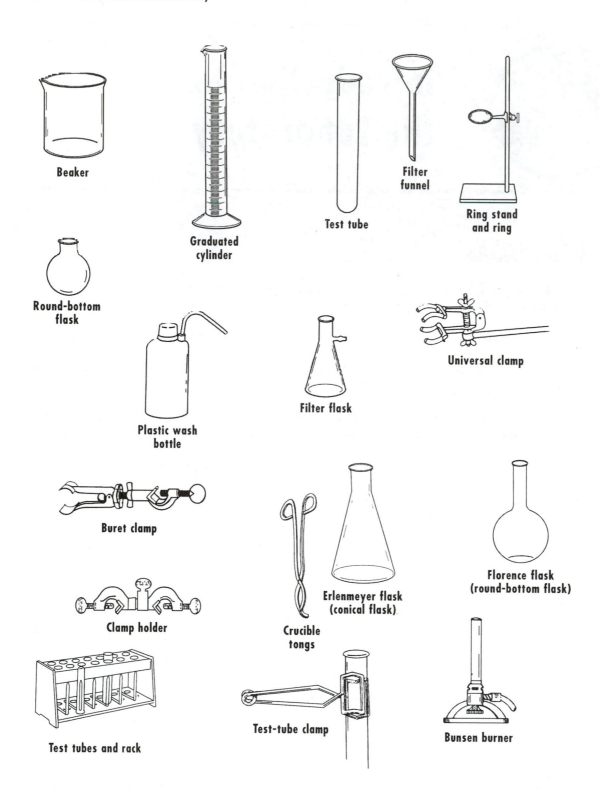

Beaker

Graduated cylinder

Test tube

Filter funnel

Ring stand and ring

Round-bottom flask

Plastic wash bottle

Filter flask

Universal clamp

Buret clamp

Clamp holder

Crucible tongs

Erlenmeyer flask (conical flask)

Florence flask (round-bottom flask)

Test tubes and rack

Test-tube clamp

Bunsen burner

Figure Intro -1

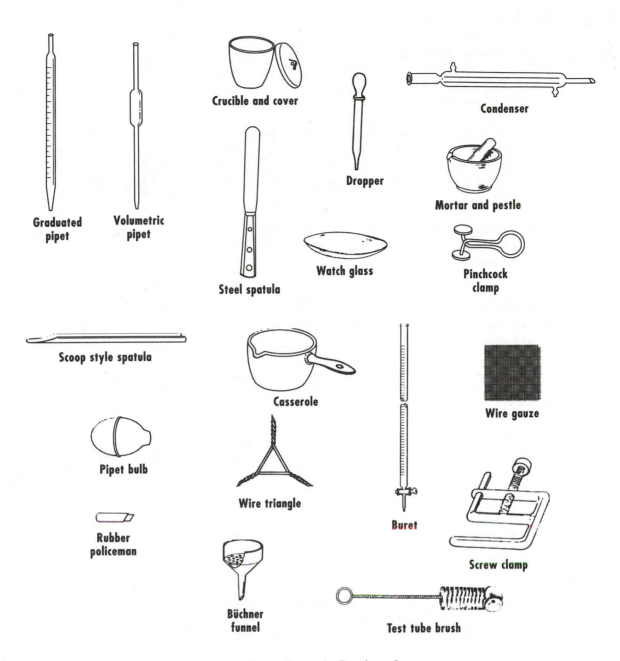

Figure Intro-1. Continued

SAFETY EQUIPMENT

I. EQUIPMENT FOR COMMON USE:

Fire Extinguisher. Students should record the location of the laboratory extinguishers and be familiar with the printed directions on each unit for its use.

Eye and Face Wash Fountains. Water has been termed the universal solvent and is the wash of choice to remove chemicals on the face and around and in the eye. A good wash fountain will provide and ample flow of aerated water and should be used for fifteen minutes when chemicals are to be removed from the face and/or eyes. If there is any doubt at all about complete removal, a physician should be consulted after the wash.

Safety Showers. A deluge-type shower will remove, or reduce to reasonably safe levels, large chemical spills on the body. Any massive exposure to corrosive or toxic chemicals should be removed in this way. Clothing should be loosened or removed to be sure the skin is thoroughly flushed.

Fire Blankets. Flaming hair or clothes can be best smothered with a safety blanket. However, the popular polyester fabrics tend to melt and smothering consequently increases the skin exposure and burn damage. The safety shower and/or carbon dioxide fire extinguisher should be employed in case of a polyester clothing fire.

Sand Buckets. A scoop of sand will extinguish most fires that might erupt in desk-top laboratory equipment. The fire extinguisher should be employed as a back-up in case flammable fluids are spread across the desk top or floor.

Ventilation Hoods. Large volumes of any gas produced in the laboratory can pose a hazard, and small. even trace amounts, of some gases can be dangerous. Proper use of a ventilation hood enables the student to do some operations involving the production of gases and vapors that would otherwise be dangerous.

Waste Containers. The waste paper baskets should not be used for the disposal of chemicals, matches, or broken glass. Many fires, explosions, and pollution can be eliminated by placing waste materials in the appropriate waste container marked for the particular kind of waste.

First Aid Kits. Relief for minor cuts and burns can be provided by your instructor at the first aid station. Also a chart for first aid procedures is provided. In every case beyond the ordinary cut or burn, professional assistance should be sought. Names and telephone numbers for on-campus and off-campus professional help in first aid advice and follow-up procedures are posted at the first aid station.

II. INDIVIDUAL SAFETY EQUIPMENT:

Safety Goggles. (see Figure Intro–2) Every laboratory worker should protect his/her eyes from chemicals with safety goggles. The goggles should be worn over prescription glasses if they are required. Contact lenses should not be worn in the laboratory as they can trap and hold chemicals against eye tissue.

Laboratory Aprons. Lab aprons will pay for themselves many times over in the preservation of clothing. Minute amounts of chemicals as solid particles or spray droplets may cause damage to clothing that will not be noticed until hours later. When working with corrosive chemicals such as strong acids and bases, a coated apron can be the difference between a chemical skin burn and a mess to be cleaned from the apron surface.

Laboratory Gloves. Lab gloves can protect hands from chemicals, burns, and minor cuts. Remember, a number of harmful chemicals can be absorbed through the skin.

LABORATORY RULES FOR SAFETY

1. Study the experimental procedures and know how to deal with the safety problems prior to entering the laboratory.

2. Laboratory work should be quiet and thoughtful. Do not interrupt other workers and never leave your experiment unattended.

3. Dress appropriately. Bulky clothes, hanging jewelry, and loose flowing hair should be avoided. Tie back long hair. Cotton clothes under suitable aprons or coats are desirable.

4. Eye protection should be worn at all times, and contact lenses should never be worn during laboratory work.

5. If chemicals come into contact with your skin or eyes, wash immediately with large amounts of water and then consult your laboratory instructor.

6. Avoid clutter on the laboratory desk; laboratory directions and equipment are all that is required. Keep all equipment and work areas clean.

7. Know the location and use of all laboratory safety equipment.

8. No food, beverage, or smoking is allowed in the chemistry laboratory.

9. Do not taste chemicals.

10. Never smell any vapor or gas directly. In those few cases where you are instructed to smell a vapor, use a cupped hand to waft a small sample toward your nose.

11. Do not attempt to insert glass tubing or thermometers into stoppers without first consulting your laboratory instructor for proper technique. Serious cuts have resulted from attempts to insert glass tubing into stoppers. It is strongly recommended that students be provided with pre-assembled apparatus when stoppers with glass tubing are required.

12. All accidents should be reported to your instructor.

13. Never work alone in the laboratory.

14. Be familiar with the first-aid procedures displayed in the laboratory.

15. Wash your hands on leaving the chemistry laboratory.

Student Commitment:

    I have read the safety rules of the chemistry laboratory and I agree to abide by them and to abide by the further safety instructions relative to each experiment that will be given to me by my laboratory instructor.

———————————   ———————————————————————

Date   Your Signature

Waft toward your nose

Figure Intro -2

SAFETY CONSIDERATION ICONS

EYE   **CLOTHING**

CUTS   **ELECTRICAL**

POISON   **BURN**

FIRE   **EXPLOSION**

HAND

LAB TEXT 1

Chemicals in Today's World: Managing Risk

In the past, "Better Living Through Chemistry," was a slogan which was not only positive but was, in fact, a philosophy which did produce better things. However, as our demand for better products increased, we soon realized that there was a price to pay in satisfying our appetite. Air emissions, waste-water, community exposure, worker exposure and hazardous waste disposal were just a few of the problems that required—and still require our attention. This heightened awareness was strengthened by a better understanding of environmental and health effects of many substances, and as our population grew, the magnitude of the problem grew.

Initially, community leaders dealt with the problems by making the word chemistry or chemical a dirty word. Questions frequently asked were, "Why even have these hazardous substances?" or "Why have industries who use and produce hazardous substances?" The fact is, everything that we have in the home or workplace is either hazardous itself or requires a hazardous material of some type in its production. The butane in a Bic lighter is as flammable and potentially explosive a substance as one can encounter anywhere. Common fertilizer was used to make the bomb that blew up the Murrah building in Oklahoma. Drano and Liquid Plumber are as caustic and corrosive as any common chemical found in a chemistry laboratory. Acid from a car battery will cause almost instant blinding, permanent scarring, and destruction of clothing and metal if spilled in the wrong place. Many young people have died from "huffing" the fumes of paint and paint removers available at local supermarkets. Virtually every prescription medication—designed to heal and cure—is deadly when taken in the wrong amounts or in combination with the wrong other materials. Chlorine, which is used to disinfect the majority of the water consumed in this country and make the water safe for consumption, is one of the most toxic gases that one can encounter, and breathing chlorine will do irreparable damage to the lungs and possibly cause death.

We must, therefore, learn to manage our hazardous materials, because as long as we members of the world community demand these products, they are here to stay. Paint on a new automobile is expected to not only look good but to last until the loan is repaid. Medication found in the home medicine cabinet has the potential to heal as well as to injure. We all demand products made from hazardous chemicals. We still want "better things" and more of them. The key is proper respect and management of the substances.

There are several approaches which have been taken to address these problems. Industries have made and continue to make great strides in reducing emissions and waste through the use of less toxic chemicals, waste minimization programs and enhanced pollution control devices. The aggressive education and training of employees as well as consumers in the safe handling of chemicals have also had a significant impact on the problem. While no single approach will solve the problems, the combination of all of these strategies strongly indicate that progress is being made.

Although progress is being made in addressing these environmental and health related areas, ultimately the question is, "Can we live in a risk-free environment?" The answer to this question may depend upon how you define risk-free. The U.S. Environmental Protection Agency's definition of "no significant risk" in the evaluation of a potential cancer causing compound, is that the risk of getting cancer as a result of exposure to a particular environmental agent is greater than 1 *increased* incidence of cancer in 100,000 people over a 70-year lifetime. When one considers that the probability of getting cancer may approximate 1 in 5 over a 70-year lifetime in the *absence* of this agent, one can understand how EPA has taken a conservative approach in their definition.

As good citizens and caretakers of our environment, we have a interest in maintaining and improving our environment. With this responsibility comes the need for us to become educated and reasonable in this very difficult endeavor.

This laboratory manual is designed to assist you in developing a better global understanding of chemicals in our world. As you proceed through this manual, there are some general issues that you should understand from a personal protection perspective, as well as from an environmental protection perspective. You will be working with some chemicals which are potentially hazardous (as you do around the house!). The Occupational Safety and Health Administration defines a hazardous chemical as, "ANY CHEMICAL WHICH IS A PHYSICAL OR HEALTH HAZARD." This definition is so broad that it's easier to say that all chemicals are potentially hazards. Water is a chemical and is a "physical or health hazard" if it is deeper than your head and you cannot swim!

There are several chemicals used in these experiments which exhibit hazardous characteristics. A typical physical hazard is flammability. This means that if the chemical is exposed to a spark or flame, it will ignite. Obviously, the key to management of the use of this chemical is to avoid having any source of ignition in the vicinity of the chemical or its fumes. Another physical hazard associated with chemicals used in these experiments is that of corrosivity. A familiar corrosive material is battery acid or sulfuric acid. A corrosive substance is a chemical that causes irreversible alterations in living tissue by chemical action at the site of contact. The major problem with a corrosive material is eye or skin contact. Eye protection and protective clothing will minimize this exposure, and care in handling the materials is the second line of defense after protective wear. The third major *type* of hazardous material is metabolic toxins. These are substances that are poisonous to ingest (swallow) or to breathe. Obviously, one takes steps to avoid breathing or ingesting these materials.

In addressing the health and safety issues one must consider the routes of exposure. One can be exposed to a health threatening chemical in one of the following ways:

1. Inhalation: There may be vapors created in the laboratory which affect you in an adverse way. Some individuals are more susceptible to these vapors than others. The proper management technique is to carry out procedures involving these materials in a fume hood—which draws the fumes away from the worker. The key response to over exposure to the fumes is go to fresh air, and see if the symptoms will go away. If they do not, then immediate first aid is sought.

2. Ingestion: This route of entry should never occur. Food or tobacco should never be brought into a laboratory. This can result in advertent exposure to the chemicals.

3. Contact: (Skin/Eye) Safety glasses should be worn in the laboratory *at all times* and personal protection such as protective clothing should be worn in the laboratory when advised. Immediate response to exposure to contact hazards always in involves washing or flushing with large quantities of water.

In summary, good laboratory practices involve blocking the route of entry or exposure to the chemical, that is don't breath it, don't eat it and don't touch it, but *manage* it to allow minimum risk. This is easy to do once you know the type of hazard associated with the chemical. Chemicals which are managed properly do not represent hazards to the individual or the environment.

WORLD OF CHEMISTRY
WORLD OF CHEMISTRY

LAB TEXT 2
Commonly Encountered Hazardous Materials and Safety Precautions in the Chemistry Laboratory

As LAB TEXT 1 indicates, many of the substances encountered in everyday life are classified as hazardous substances. Almost all of the substances used in a chemistry lab would be classified as hazardous materials under some categories. It would be nice to work in a lab (academic or otherwise) where the materials used posed no hazards, but this is not only impossible but would not provide realistic or useful experiences if it could be done. The answer to working safely with hazardous materials is to use proper precautions, follow directions carefully, and use common sense. Very rewarding insights and experiences can result from carefully doing the experiments in this manual, and this *can* be safely done.

In any chemistry laboratory, the same common substances are used again and again, and the same *types* of situations arise again and again. The purpose of this section of the manual is to familiarize you with a relatively small number of chemicals or families of chemicals that are used in a number of experiments so that you will know their characteristics and how to safely use them.

CLASSES OF HAZARDOUS MATERIALS

a. Acids and bases. PROBABLY THE MOST FREQUENTLY ENCOUNTERED CLASS OF HAZARDOUS MATERIALS IN THE LABORATORY ARE ACIDS AND BASES. WHEN EITHER IS SPILLED, WHETHER ON THE FLOOR, ON CLOTHES, OR ON SOFT TISSUES (EYES, MOUTH, ETC.) THE AREA SHOULD BE WASHED WITH LARGE AMOUNTS OF WATER. TRYING TO NEUTRALIZE AN ACID WITH A BASE AT A SPILL SITE IS GENERALLY HAZARDOUS—PARTICULARLY ON THE SKIN. NEUTRALIZATION PRODUCES GREAT AMOUNTS OF HEAT. WHEN ACID OR BASE IS SPILLED ON THE SKIN OR CLOTHING, FLUSHING WITH LARGE QUANTITIES OF WATER, FOLLOWED BY WASHING WITH A MILD SOAP IS THE BEST TREATMENT. A CONCENTRATED ACID SPILL *ON THE FLOOR* CAN BE TREATED WITH SOLID SODIUM BICARBONATE (BAKING SODA—$NaHCO_3$) or SODIUM CARBONATE (Na_2CO_3).

b. Flammable liquids. Open flames or sparks should be kept away from hydrocarbons, ethers, alcohols, ketones, and many other organic liquids. Some of the common volatile flammable liquids you may encounter in this course are acetone, diethyl ether, ethanol, and methanol (also known as ethyl and methyl alcohol). The word volatile means "easily evaporated," and some of these liquids are so volatile and so flammable that an open container of the liquid is a hazard if there is a flame (bunsen burner, etc.) *anywhere in the lab*. Volatile, flammable liquids should be heated only with flameless electric heaters, and should be heated in a fume hood to prevent vapors from permeating the laboratory. See the section on combustion-related explosions, below.

c. Powerful oxidizing agents. Compounds containing metals or non-metals in a very high oxidation state (very positive) are often quite hazardous by being very sensitive to heat and/or shock. Usually these involve anions with 3 or more oxygen atoms as a part of the ion. Examples are nitrates

(NO_3^-), perchlorates (ClO_4^-), chlorates (ClO_3^-). Some are not particularly shock sensitive, but are such strong oxidizing agents that if they are inadvertently mixed with an easily oxidizable material an explosion may occur. Examples of these are permanganate (MnO_4^-), chromate (CrO_4^{2-}) and dichromate ($Cr_2O_7^{2-}$).

c. metabolic poisons. Some materials are dangerous simply because they are poisonous. The gases carbon monoxide and hydrogen cyanide mentioned below are examples. Compounds containing metal ions are frequently metabolic poisons, when ingested in large amounts. The dangers of lead and mercury are discussed in the newspapers on a regular basis, and both of these can indeed cause major problems if ingested, even in fairly small amounts. Cadmium compounds are almost equally poisonous, and even nickel, copper, zinc, and other common metal compounds are toxic when their compounds are ingested. (It must be pointed out that the metals themselves, as the free metals, have no toxicity to speak of, with the exception of liquid mercury, which can be absorbed through the skin.) The actions of these poisons are usually in interrupting the action of biological catalysts called enzymes which govern many bodily processes and functions. *NO CHEMICALS ENCOUNTERED IN THE LABORATORY SHOULD EVER BE TASTED, EATEN OR OTHERWISE INGESTED. IF THIS OCCURS, NOTIFY YOUR LABORATORY INSTRUCTOR AT ONCE.*

d. corrosive agents. Many materials encountered in the laboratory are hazardous because they break down other substances (or tissue) in a corrosive fashion. Most of these are either acids or bases. The damage can be a case of actual bond breaking in the substrate being attacked (clothes, tissue, or inanimate materials such as paper or metals) or dehydration.

EXPLOSIONS

While not the *most frequent* causes of injuries in the lab, explosions frequently produce the most severe injuries. Explosions most frequently arise from one of two sources: Combustion related and oxidation-reduction accidents. Combustion related explosions occur when the lab atmosphere becomes permeated with highly flammable gases or with a large amount of combustible, finely divided solid particles (coal dust, flour, or even sawdust), mixed with the oxygen of the air. A flame, or even a spark from flipping the off/on switch of an electrical appliance may trigger the explosion.

Oxidation-reduction reactions produce the most violent explosions (gunpowder, dynamite, TNT, are this type). These occur when a strong oxidizing agent (electron acceptor) is combined with a reducing agent (electron source) and allowed to react in an uncontrolled manner. These reactions generally are highly exothermic, and occur so quickly that the great heat liberated does not have time to dissipate, and it vaporizes the products instantaneously. The destructive force of the explosion is a consequence of the rapidly expanding gases.

HAZARDOUS SUBSTANCES

a. Sodium hydroxide, NaOH. Sodium hydroxide (caustic, lye) is a very strong base and is extremely corrosive. Even 1% solutions can produce rapid and serious eye damage, and hot solution will cause instantaneous tissue destruction. Any solution with pH exceeding 10 can cause skin irritation, and concentrated solutions or the pure solid can cause severe burns upon exposure longer than a few seconds. Solutions must be prepared cautiously since the dissolving process is very exothermic and boiling and spattering may result. Spilled solution should be quickly flushed with water, and exposed skin areas should be washed quickly with lots of water. Clothes should be quickly and thoroughly rinsed.

b. Potassium hydroxide, KOH. This is a close "cousin" to NaOH (above), and has practically identical properties, with the exception that it has a significantly lower heat of solution than sodium hydroxide. It is sometimes used for a glassware cleaning bath, and the same cautions apply as with NaOH. Glassware should not be left for long periods in either of these strong bases, since these substances will etch glass.

c. Ammonia, NH_3. Ammonia is a gas with a pungent, irritating odor which is very soluble in water and which dissolves to give a basic solution. (A solution of ammonia in water is frequently called ammonium hydroxide, although a much better name is aqueous ammonia.) Both the gas and the solution are very damaging to soft tissues—particularly the eyes. The vapors from concentrated ammonia should not be inhaled nor allowed to get in the eyes. Concentrated solutions containing ammonia (concentrated ammonium hydroxide) should be used *only in* fume hoods, where the odors cannot circulate into the room.

d. Nitric acid, HNO_3. Nitric acid combines the corrosiveness of a strong acid with the destructiveness of a powerful oxidizing agent. Reactions involving HNO_3 can produce very toxic nitrogen oxides, especially NO, NO_2, and N_2O_5, so the fumes from nitric acid spills are hazardous. Solutions of 5% or greater concentration quickly destroy eye or skin tissues. Immediate washing with large quantities of water is necessary to avoid serious injury. If nitric acid contacts the skin, the spot will slowly turn dark yellow and will remain yellow for several days. This is not dangerous, and is, in fact, a test for the presence of protein. Nitric acid reacts with proteins to produce a yellow color (this is called the zanthoproteic test).

e. Phosphoric acid, H_3PO_4. This is commonly referred to as a strong mineral acid, although in truth it is not as strong as the other common mineral acids (HCl, H_2SO_4, $HClO_4$, HNO_3). It has a somewhat lower heat of dilution, but caution is still necessary when using it. It has a very low toxicity—one of the main commercial use for phosphoric acid is as a flavor additive for foods.

f. Sulfuric acid, H_2SO_4. In this case, hazards result from the combined effect of a very strong acid and a strong dehydrating agent. Solutions with a concentration below 5% are hazardous principally to the eyes. More concentrated solutions are very destructive. Concentrated H_2SO_4 produces severe burns, due to the great affinity of the acid for water and the heat evolved during solvation (combination with water). The familiar demonstration of the action of H_2SO_4 upon sugar should serve as a warning of the danger in H_2SO_4. The vapors from hot sulfuric acid will produce H_2SO_4 in the lungs. Evaporation of H_2SO_4 solutions should be carried out only in the hoods. Sulfuric acid spills should *immediately* be cleaned up with copious amounts of water.

Dilution of sulfuric acid must be carried out cautiously in order to avoid excessive heating and spattering. Since H_2SO_4 is more dense than water, the acid should be poured slowly into water with stirring.

g. Chromic acid. A solution of CrO_3, K_2CrO_4 in concentrated H_2SO_4 has been used frequently in the past as a solution for cleaning laboratory glassware. The qualities that made it a good cleaning agent also make it hazardous. It combines the dangers of sulfuric acid with the dangers of a strong oxidizing agent. Its reactions with organic compounds may result in explosion. The extreme toxicity of Cr(+6) to aquatic organisms has greatly curtailed the use of chromic acid as a laboratory reagent, due to the great difficulty in safely disposing of this material.

h. Hydrochloric acid, $HCl_{(aq)}$. Concentrated solutions of Hydrochloric acid (hydrogen chloride gas dissolved in water) are quite corrosive, and gaseous HCl readily escapes, creating quite hazardous fumes from the liquid. Dilute solutions are not as corrosive, but should be washed off immediately with water. Even dilute solutions of hydrochloric acid are very damaging to eye tissues. The fumes from the acid itself *or from its reactions* must not be inhaled.

i. Perchloric acid, HClO$_4$. Perchloric acid is an important laboratory reagent. It is hazardous for several reasons:

(i) It is one of the strongest acids used in the lab.

(ii) Fumes from hot perchloric acid and the fumes from its reactions are highly toxic.

(iii) Perchloric acid, perchlorates, and many derivatives of these are very explosive. (see "explosives" below)

j. Carbon monoxide, CO. Carbon monoxide combines with the hemoglobin in blood and deprives it of its oxygen-carrying capacity. It is produced whenever carbon-containing material burns in an environment where insufficient oxygen is present to fully support the combustion. Small concentrations can be deadly. Sub-lethal doses can lead to severe headaches, which should serve as warnings to seek fresh air, except that the associated grogginess may prevent clear thinking. Laboratories should be well-vented to avoid buildup of gases such as carbon monoxide.

k. Carbon dioxide, CO$_2$. This is a colorless and odorless gas which is not toxic when inhaled, but which can produce suffocation through oxygen deprivation if allowed to accumulate in large quantities. This is not a problem in labs with normal ventilation systems.

l. Hydrogen chloride, HCl. This is a colorless gas which frequently appears white or light gray due to interaction with water vapor in humid air to form microscopic drops of hydrochloric acid (see "h" above). It has a very pungent, biting odor and can cause severe damage to lung tissue through the formation of hydrochloric acid on the inner linings of the lungs and respiratory track. It is formed when strong mineral acids (H$_2$SO$_4$, H$_3$PO$_4$) react with metal chlorides (NaCl, KCl). If HCl gas is present in large quantities in a laboratory atmosphere, the laboratory should be evacuated immediately.

m. Hydrogen sulfide, H$_2$S. This is another extremely poisonous gas. Its odor (like rotten eggs) provides a warning, but the olfactory organs can become desensitized. The human nose is so sensitive to hydrogen sulfide that concentrations far below a dangerous level can be detected, and in fact laboratories with very small quantities of hydrogen sulfide gas in the air can be quite safe but very unpleasant places to be!

n. Sulfuric dioxide, SO$_2$. Sulfur dioxide is produced upon burning sulfur or almost any sulfur compounds. Its irritating odor is an incentive not to breathe it, and it is highly toxic.

o. Bromine, Br$_2$. Bromine is a very reactive liquid element, and it is dangerous on external contact or upon breathing the vapors from the liquid. (*This warning applies also to the other halogens*). Liquid bromine produces deep and long-lasting burns, and should be handled with a great deal of caution.

p. Phenol, carbolic acid, C$_6$H$_5$OH. Phenol produces severe burns and is a serious hazard to the eyes or skin. It also is absorbed through the skin and produces acute systemic poisoning in this way.

q. Hydrogen cyanide, HCN. This is a *highly toxic* gas with the odor of almonds. It is produced whenever a cyanide salt (NaCN, KCN, etc.) comes in contact with a mineral acid (H$_2$SO$_4$, H$_3$PO$_4$, etc). *NEVER STAY IN A LAB WITH HCN VAPOR PRESENT FOR A MOMENT!*

r. Nitric oxide, NO. This is a colorless but quite pungent gas produced mainly as a product from internal combustion engines (when N$_2$ and O$_2$ in the air react during the explosions of the gasoline/air mixture in the cylinders). It is also produced by decomposition of some nitrogen-containing compounds in the lab, but is not commonly a problem in laboratories.

s. Nitrogen dioxide, NO_2. This gas has a rich brown color and a very pungent odor akin to that of chlorine gas (Cl_2). It is produced by the action of nitric acid on certain (reasonably unreactive) metals such as copper, as well as by air oxidation of nitric oxide. Photochemical smog in cities is "dirty brown" as seen from a distance, due to the presence of NO_2 as a pollutant. It is toxic, and reactions which produce NO_2 should be carried out in a fume hood.

t. Sulfur Trioxide, SO_3. This is a white (NOT invisible!) gas which is produced upon prolonged heating of concentrated sulfuric acid (H_2SO_4). Commercially available **fuming sulfuric acid** contains about 30% SO_3 dissolved in 100% sulfuric acid, and is an *exceptionally* strong hazardous acid. If SO_3 vapor is inhaled, it instantly becomes sulfuric acid on the lining of the lungs an respiratory tract and causes severe damage. Use of fuming sulfuric acid or strong heating of concentrated sulfuric acid MUST ALWAYS BE DONE IN FUME HOOD!

CLEANUP

A reasonably effective and easily prepared medium for cleaning up a variety of spills can be prepared by mixing equal parts of sodium carbonate (caustic soda), kitty litter, and charcoal (with briquettes crushed up a little). This mixture can be poured on a liquid spill, or a liquid can be poured into a container of the mixture. The sodium carbonate neutralizes acids; the kitty litter absorbs and partially neutralizes bases, and the charcoal and kitty litter both absorb organic toxicants. If enough of the mixture is added to absorb all of the liquid and still remain solid, the material can safely be disposed of as a solid.

Solutions containing transition metal ions should not be flushed down the drain in the laboratory. This is because these are metabolic poisons for bacteria, fish, and animals just as discussed above for humans. These should be discarded in a large collection vessel maintained just for that purpose. The metals can be safely removed from solution after collection by precipitation (converting them into a water-insoluble compound). This is easily accomplished by adding a source of sulfide ions to the solution. The metals precipitate as the sulfide salts, which are so very insoluble that they are effectively inert and can safely be landfilled. Indeed, many natural metal ores, which have been in the ground for centuries, are sulfide compounds (galena (lead ore), cinnabar (mercury ore), and pyrite (iron ore)).

Acid and base solutions are used in the lab are usually fairly dilute and can safely be washed down the drain with lots of water, but this *SHOULD NEVER BE DONE WITHOUT THE KNOWLEDGE OF YOUR INSTRUCTOR!* Concentrated acid or base is usually diluted first, then neutralized prior to disposal.

LAB TEXT 3
The Laboratory Notebook

One of the most indispensable resources in the chemistry laboratory is the laboratory notebook. It serves as the first hand record of all operations, results, observations and impressions that take place or arise in the lab. It contains all measured data, a chronology of how the data was obtained, descriptions of equipment and procedures used, and the conjectures and conclusions of the individuals doing the work. The importance of maintaining an accurate record of events in the laboratory cannot be overemphasized. A laboratory notebook can provide the blueprint for repeating an experiment that has produced an exciting or unexpected result. It provides a roadmap to discover reasons why an experiment has not gone as expected. It can serve as legal evidence in a patent dispute or simply a guide for the next person working on the problem to follow. The notebook insures one can learn from his/her own mistakes or those of others, and that one can use previous successes on which to build future efforts.

The obvious question is, "What goes into the laboratory notebook?" The obvious answer is, EVERYTHING! It is not an uncommon posture among companies that rely on their research laboratories for product development to have a corporate policy that says, "If it is not written in the laboratory notebook, then it did not happen in the lab." Thus, if a laboratory worker should make an exciting new discovery, *the discovery has no status whatsoever if the research has not been documented in the notebook*! This may seem like a harsh or even self-defeating policy, but in fact it is exactly the opposite. In order for benefit to be derived from anything occurring in the laboratory, it must be possible to both understand exactly what transpired, and to exactly duplicate the work on demand. This can only be done if accurate records of the proceedings exist.

Laboratory notebooks may be maintained in a variety of styles, but all styles should include certain common elements. These are as follows:

(a) *Pagination.* Pages in a laboratory should be numbered consecutively. This allows the development of a table of contents at the beginning of the notebook, which provides for quick reference when specific procedures or data need to be recovered.

(b) *Date.* Each time that work is done, the work should be dated. Dated laboratory notebooks will serve as evidence of primacy in patent disputes and serve to clearly illustrate the chronology of a research project.

(c) *Clear title or descriptive heading:* Each time a new experiment or project is begun, that section of the laboratory notebook should be headed by a clear title or descriptive statement that clearly identifies the purpose of the work being recorded.

(d) *Identification of individual(s).* When a notebook is serving as the record for a group working on a project, the individuals responsible for each bit of work should be identified by initials or names, for later reference purposes.

(e) *Unusual conditions existing in the lab.* If anything is out of the ordinary on a particular day, the unusual conditions should be recorded, as this might be a reason for an unusual or unexpected result. For example, if the air conditioning is not working on a hot summer day, the elevated temperature will likely cause processes to take place more rapidly, and higher humidity might introduce a moisture factor into a reaction that would not be a problem in an air conditioned lab.

(f) *Descriptions of any apparatus or equipment.* Any commercial instrument used in the lab should be identified by name, manufacturer, and model number. A special apparatus assembled from laboratory equipment should be described clearly—with sketches where appropriate. Glassware such as beakers, flasks, etc. used should be identified (ex. "a 400 mL beaker was partially filled with . . ."). The device used for a particular measurement should be identified (ex. "20 mL of this solution was pipetted . . .").

(g) *All numerical data, including identifying information and units.* Any measured number taken in the lab should be recorded in the laboratory manual. Indeed, NO PAPER OTHER THAN THE LAB MANUAL SHOULD EVER BE USED FOR RECORDING DATA! The following scenario illustrates the reason for this policy. Suppose an individual weighs a beaker, and writes the weight on a scrap of paper by the balance. A reactant is added to the beaker, and the weight is again measured. The weight of the empty beaker is then subtracted from this final weight to obtain the weight of reactant used, and the net weight of the reactant is recorded in the lab manual, while the piece of scrap paper with the weights of the beaker and beaker plus reactant is discarded. If a simple error in subtraction had been made before the number was recorded in the laboratory manual, there is now *no way to ever recover the correct information*! The entire experiment has been rendered useless by a careless error, whereas, if the actual weights had been recorded, they could be checked for accuracy at any time in the future. The same principle applies if a volume, determined by the difference of an initial and final reading, is recorded, rather than the readings themselves. ALL MEASURED NUMBERS SHOULD BE RECORDED—EVEN WHEN A DIFFERENT NUMBER, RESULTING FROM A COMBINATION OF THE MEASURED NUMBER IS ACTUALLY SOUGHT. Many laboratories are now equipped with modern balances which automatically subtract the weight of the container for the person doing the measuring. Typically, one weighs the empty container, then sets the balance to zero *with the container on the balance pan*. The weight of the material added to the container can then be read directly from the balance, since the balance has electronically subtracted the weight of the container. The net weight of the material can then be recorded directly in the notebook. When a balance with this capability is used, the word "tared" should be written beside the net weight of the substance to indicate that the subtraction was electronically done, and that there is no missing data such as the weight of the empty container.

(h) *A description of the procedure used, or a reference to where the description may be found.* If the procedure being used is recorded in a published source, such as a research paper or a laboratory manual, one needs only to reference the source, *while identifying any changes made from the published procedure*. For example, one might say "the procedure used for the analysis was that described in LABORATORY MANUAL FOR CHEMISTRY IN THE MODERN WORLD, page 97 (Gravimetric Determination of Per Cent Sulfate in an Unknown Compound) with the exception that a solution of $Ba(NO_3)_2$ was used, rather than $BaCl_2$." If the procedure is not from a published source, then it should be described in the notebook. One does not have to describe "standard operations" in detail, but just mentions them by name. For example, one would say "the solution was titrated against a standard solution of NaOH" rather than describing the process of titrating.

(i) *Qualitative observations of occurrences.* When observable changes occur during a procedure, they should be described, such as "the dissolving of the copper was accompanied by the evolution of a dark reddish-brown gas with a pungent odor" or "the solution changed from a light sky blue to a dark blue color upon the addition of the 15 mL portion of concentrated ammonia."

In summary, the laboratory notebook is your record of the happenings in the laboratory, and should contain all information necessary to describe clearly (1) what you did, (2) how you did it, and (3) what results you obtained.

You should maintain your own laboratory notebook and include in it each experiment done in the class. This laboratory manual contains data pages at the end of each experiment to facilitate the reporting of your results at the end of the proceedings. These data pages are not meant to replace the laboratory notebook. All initial data should be recorded in your laboratory notebook, and the data pages at the end of the experiments should be filled out using the data in the laboratory notebook.

LAB TEXT 4
Preparing and Using Graphs

Graphs are visual indicators of the relationship between two varying quantities that depend upon one another. They generally show the relationship between an *independent* variable and a *dependent* variable. As the names imply, the value for the dependent variable is a function of (depends on) the value of the independent variable.

Consider a family taking a trip in an automobile. To monitor their progress on the trip, they might choose to record the distance they have traveled as a function of time, and record the odometer reading (mileage) each half-hour. The distance traveled, as registered on the odometer, is the dependent variable and time is the independent variable (just try and make time depend on anything!). If the family is traveling on an uncrowded interstate highway with the cruise control feature in operation, then the distance traveled will be almost exactly proportional to the time elapsed, and the collected data will look like that in Table LT4-1. The graph of this data, shown in Figure LT4-1, will be linear, with the slope of the line showing the relationship between the variables distance and time. This is the speed, which is reported in miles per hour.

Time (hours)	Distance (miles)
0.5	33.2
1.0	66.1
1.5	98.9
2.0	136.4
2.5	167.2
3.0	198.3
3.5	231.0
4.0	265.1

Table LT4-1

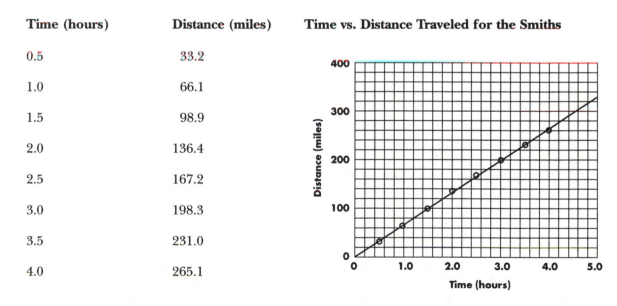

Figure LT4-1. Smith Family Travel

Suppose instead that the Smith family had chosen to travel on a train which has to go up a long mountain, and loses speed on the way up, before gaining speed on the way down. The slope of the line at any point will still provide the average velocity at any particular time, but the average velocity will change and the graph will not be a straight line as a result.

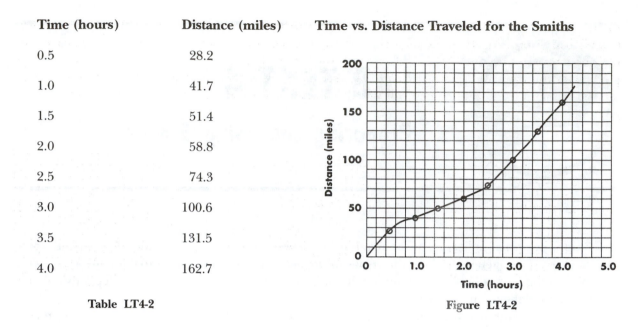

Time (hours)	Distance (miles)
0.5	28.2
1.0	41.7
1.5	51.4
2.0	58.8
2.5	74.3
3.0	100.6
3.5	131.5
4.0	162.7

Table LT4-2

Figure LT4-2

Linear graphs, like the one in Figure LT4-1, frequently provide more useful information than non-linear graphs like the one in Figure LT4-2. For example, one could easily predict that the Smith family would travel about 330 miles in 5 hours on the interstate, using the slope of that graph (speed equals 66 miles per hour). The prediction for the train journey would be much less certain, since the speed (slope) is not nearly as constant.

When graphs are prepared from tabular data, there are some important principles that must be observed. The first is that a well-presented graph should contain all of the information to indicate exactly what the graph depicts, and the actual numbers being used. This is accomplished by *heading the graph with a clear title* that shows the variables being related to one another (note the titles for the graphs in Figures LT4-1 and LT4-2), *clearly labeling the axes,* showing the units used for the measurements (distance in miles and elapsed time in hours on the above examples), and *clearly indicating the numerical values on the axes.*

The second principle is that the plotted data points should be spread out over as much of the graph as is possible. This is accomplished by using appropriate *scale factors* for each axis. The scale factors are the numerical increments represented by each line on the graph. For example, appropriate scale factors can be estimated by dividing the limits of the range of data (4.0 hours and ≈275 miles in Table LT4-1) by the number of lines available on that particular axis of the graph paper to be used in preparing the graph. For the data in Figure LT4-1, this produces scale factors of 16.0 hours/line and 13.75 miles/line for the X and Y axes, respectively. The scale factors are rounded off to the nearest "convenient" numbers so that the graph can be laid out in the most efficient manner for plotting the points. You will find that convenient scale factors are easily divisible by 2, 4, 5, or 10. Scale factors divisible by 3 generally make plotting and using data from graphs much more difficult and should be avoided if possible. Figure LT4-3 shows the data from Table LT4-1 plotted with three different sets of *inappropriate* scale factors. In (a), the X scale factor is too large, in (b) the Y scale factor is too large, and in (c), both scale factors are too large. In each case, the resulting graph has the plotted data concentrated in too small a region. This limits the precision with which the graph can be read and interpreted.

A third important principle involves the drawing of the line or curve. A popular children's exercise is playing "dot-to-dot" where dots on a page are connected in order to show a tiger hiding in the grass or a bird in a tree. Many times a graph is *incorrectly* drawn in the same fashion. In order to understand how a graph should be drawn, one must remember what a graph represents. The graph shows the relationship between two variables, and involves measured quantities. These relationships almost invariably involve very regular patterns, which may or not be linear. Those that are non-linear are almost always *smooth* curves. Since all measured quantities have inherent uncertainty in the measurements, the plotted data usually does not provide an exact or perfect fit to the theoretical relationship, but each individual point should be fairly to very close to the actual line or curve. What insight does this discussion provide about the method of fitting the data to the best line or curve? The answer is, the line or curve *should not be forced through any single point*, but rather should be drawn so that it approximates the pattern displayed by *all* of the data points. A rather striking result of this principle is that a line might be drawn that best suits the overall data, but which does not go through a single data point!

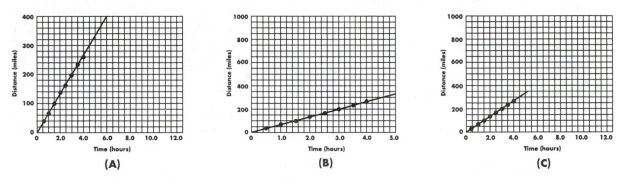

Figure LT4-3. Graphs with inappropriate scale factors.

Consider the problem of determining the average weight of 35 students in a class. One would weigh each student, total the weights, and divide the total by 35 to get the average weight. Such an average, for college students, could turn out to be 188 lbs. There might not be any students in the class that actually weigh 188 pounds, but this figure would be a very good one to use if 15 of the students were going to be on a large boat and the captain needed to know about how much the group would weigh (15 × 188 = 2820 pounds). The average value is the number of choice when a large number of students is to be considered. So in a graph is the *slope* the value of choice to relate one variable to another. The slope represents the *average* measured relationship between the dependent and independent variables.

Consider now an experiment designed to measure the amount of light absorbed by a solution. The absorbance of a solution is known to be directly proportional to the concentration of the light-absorbing species that is dissolved in the solution—that is, the relationship between absorbance and concentration is a linear one. Several solutions are prepared, with different concentrations, and the absorbance of each solution is measured (see experiment 29 for a discussion of this). The concentration vs absorbance data is then plotted, and the data does not produce a "perfect" line, even though the relationship is known to be a linear one. This is because there is natural uncertainty in determining both the concentration and the absorbance, so any given measured data point might not fit exactly on the "true" line, although it will likely be very close. How, then, does one find the best line if no particular point can be known with absolute assurance to lie on the line? The answer is to use a straight edge and draw *one straight line across the graph that comes as close as possible to ALL data points!* It may or may not go through one or more data points, but it should be very close to most or all of the points. Figure LT4-4 shows two graphs of the same data. In (a), the "line" has been *INCORRECTLY* forced through all data points, and in (b) the line has been *CORRECTLY* drawn to approximate all data points. Note that in each case the data points have been accurately plotted, and a small circle has been drawn around each. The circle indicates the inherent uncertainty in the measurement, and the line is drawn by looking at the circles. This is a common procedure and a good one to use.

If the relationship is known to be a non-linear relationship, or if it is unclear whether it is linear or not, then the procedure is to draw a *smooth curve* that comes as close as possible to all of the points without producing humps or angles. Few natural variables produce sharp turns or angles in an otherwise gradually changing relationship, even if the relationship is not a linear one.

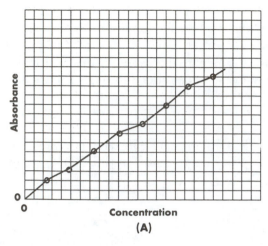
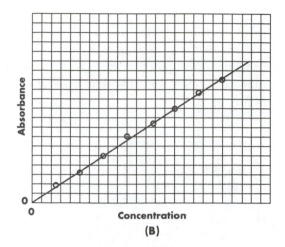

Figure LT4-4. Incorrectly and correctly drawing a line.

For either linear or non-linear graphs, once the line or curve is drawn, *the individual data points have no further status whatsoever!* this means that the line is the "true" relationship and the measured data points are ignored for all further use of the graph. The drawing of the line or curve actually serves to provide a best average of the data points in determining the actual relationship between the variables.

A fourth important principle involves the graph paper itself. The usefulness of a graph is often determined by the quality of the graph paper chosen for the graph. Some graph paper has large squares, and some has very small squares. As a general rule, *the smaller the subdivisions are, the more precisely information may be plotted and recovered from graphs.* Graphs should be done on graph paper with the smallest available subdivisions. This allows for precise plotting of data points, and information to be read from the graph can be determined with greater precision. Paper with squares that are as large as 1/4 in on a side is usually unacceptable for scientific work, and paper lined in 1/10 inch or 1.0 mm subdivisions is the preferred paper.

USE OF THE GRAPH — CALCULATIONS

For linear graphs, the slope of the line is usually of paramount importance. The slope is a measure of the rate of change of the dependent variable with respect to how fast the dependent variable is changing. It is calculated by identifying two points on the line and dividing the change in the y variable by the change in the x variable

$$(\frac{\Delta y}{\Delta x} \text{—sometimes called "rise over run").}$$

Three important principles must be observed when calculating the slope. The first is that *data points used to draw the line must not be used in calculating the slope.* The correct procedure is to mark two points from the line, read their coordinates (x_1, y_1 and x_2, y_2) from the respective axes, and to calculate the slope $(y_2 - y_1)/(x_2 - x_1)$. The second principle is that *the points chosen for calculating the slope should come from opposite ends of the line, rather than points close together.* This will make the values of Δy and Δx as large as possible and as a result will minimize any uncertainty in identifying the value of the coordinates themselves. The third principle is that the coordinates should be read *to the limit of precision that the scale actors on the axes allow.* The axes represent scales, and the technique for reading values from these scales are the same as the techniques for reading scales on measuring devices, as described in Experiment 1 of this manual. The scale is read to one more place than the actual graduations, with the last place being the best estimate between the lines (see discussion on reading scales in Experiment 1).

For non-linear graphs, the slope at any particular point is frequently a valuable parameter as well. In this case, however, the first step is to draw a line tangent to the curve at the point where the slope is desired. The slope of this tangent line is then determined using the principles outlined above.

A frequent use of a graph is to use the relationship provided by the line or curve to predict values of one variable when values of the other variable are known. This is done by locating the known value on the appropriate axis, then drawing a right angle through the line to find the corresponding value for the other variable from the other axis. If the value to be estimated lies within the range of the measured data points used to draw the line, the process is known as interpolation (prefix inter = between) whereas if the value to be measured lies outside the measured range of data points the estimation comes from a region of the graph produced by extending the line and is known as extrapolation (prefix extra = outside). Using the graph of the Smith family's interstate travel (Figure LT4-1), one can interpolate that the distance traveled in 3.25 hours was about 215 miles, and extrapolate that the distance that would be traveled in 4.30 hours would be about 285 miles. You should verify these estimates for yourself.

1 Introduction to Laboratory Measurements

OBJECTIVE

To become familiar with common measuring devices used in a laboratory; to learn the sensitivity with which these devices can be used; and to gain an appreciation for the degree of uncertainty associated with data obtained in the lab.

APPARATUS AND CHEMICALS

burets with stands	metal object
100-mL graduated cylinder	thermometers
10-mL graduated cylinder	calculator
50-mL beaker	burner
400-mL beaker	ring
copper wire or paperclip	ring stand
tongs	wire gauze
meter stick	clamp
balance	one-hole, split rubber stopper

SAFETY CONSIDERATIONS

 Use care in handling the graduated cylinder. While laboratory glassware is made of durable glass, it can shatter if mishandled, and can cause severe cuts.

 No immediate eye danger is present in this particular experiment, but as a part of your introduction to the lab you should be made aware that *safety glasses are to be worn at all times* in the lab.

 Mercury is poisonous and can be absorbed through the skin. A broken thermometer can splatter mercury in the working area.

 When using a burner or any open flame in the laboratory, be absolutely sure that loose hair and clothing do not come close to the flame. Many hair sprays contain highly flammable components that cause hair to burn quickly. Long hair should be tightly tied back in the lab. It is easy to forget these simple rules when concentrating on measurements.

 Be sure the gas is turned off completely when you are through with it since in a closed laboratory a small gas leak can build up to an explosive mixture.

FACTS TO KNOW

Much of what is done in a chemistry lab involves taking measurements on a system using appropriate measuring devices, and then making calculations with the numbers obtained. Numbers used in such calculations fall into two broad categories: pure numbers and estimated numbers. Pure numbers are those with no uncertainty at all. These include small numbers of *counted* items (the number of eggs in a nest or puppies in a particular litter, for example) and *defined* numbers (3 feet in 1 yard, 16 ounces in 1 pound, for example). In each case, the numbers are generally written as whole numbers, but could be written correctly to any number of places (3 feet = 1 yard, or 3.00 feet = 1.00 yard). It is the other category of numbers—estimated (measured) numbers—with which chemists deal most frequently.

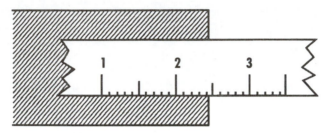

Figure 1-1. Estimation of length as 2.43.

Most measuring devices have some sort of scale to be read. The scales will have individual marks, called graduation marks, and will have printed numbers at regular positions along the marks. The proper way to use such a scale is to **record the measured value to one more place than the scale is marked.** For instance, if using a meter stick that is marked in millimeters (mm), record the length to 0.1 mm. With a cylinder graduated in milliliters (mL), record the value to 0.1 mL. The last place in the number is **your best estimate** of the fractional part of the graduation. For example, in Figure 1-1, it can easily be seen that the value of the reading falls between 2 and 3, and, by counting graduations, that it falls between 2.4 and 2.5. One can *estimate* that the reading falls about 3/10 of the way between the finest graduations, giving a measurement of 2.43. The last number *could* be a 2 or a 4, but the measurement is certainly much closer to 2.43 than to 2.4 or 2.5. The best way to record such a reading would be 2.43 ± .01 to show reasonable limits of uncertainty. In this case, the actual value is probably within 0.5% of the measured value, and this can be easily seen by anyone who examines the number. When calculations must be done with numbers obtained through measurements taken in a laboratory, it is very important to know the degree of uncertainty in the measured values. This is done through the use of **significant digits,** that is, figures in which the uncertainty is only in the *last recorded digit* (for example, the 3 in the recorded number 2.43 discussed above).

EXAMPLE 1

Fourteen cherries are placed in a pre-weighed bowl and found to have a total weight of 21.17 grams. What is the average weight of a cherry, to the correct number of significant digits?

Answer: The 14 is a pure or exact number and has an unlimited number of significant figures. The weight is a measured number and contains uncertainty (estimate) in the fourth significant digit. Thus, the answer will have the same degree of uncertainty as the *least* certain of the numbers being divided, or: (21.17 grams)/(14 cherries) = 1.512 grams/cherry. (Note, the position of the decimal *does not matter* as far as the number of significant digits is concerned.)

When multiplying or dividing groups of measured values the answer will have the same number of significant digits as the least accurately known (fewest significant digits) of the numbers being multiplied or divided.

EXAMPLE 2

Using a balance, a student determines the mass of an object to be 12.37 ± 0.02 grams. She then uses a graduated cylinder and measures the object's volume to be 3.12 ± 0.02 cm^3. What is the density of the object, reported to the correct number of significant figures, and how could one show WHY there is a "correct" number of significant figures?

Answer: The object's density, calculated using these numbers *and ignoring significant figures* is 3.96474 g/cm^3. The ±0.02 numbers represent reasonable uncertainty in reading the scales. Thus, while the *best estimate* for the mass was 12.37 g, it could reasonably have any value between 12.35 g and 12.39 g, while the volume could reasonably have any value between 3.10 cm^3 and 3.14 cm^3.
(a) Treating the *mass as a pure number* (no uncertainty) and the volume as a measured number produces a range of densities from 3.9395 g/cm^3 to 3.9903 g/cm^3 (dividing 12.37 by 3.14, then dividing 12.37 by 3.10). This is a variation of 0.0508 g/cm^3.
(b) Treating the *volume as a pure number* (no uncertainty) and the mass as a measured number produces a range of densities from 3.9583 g/cm^3 to 3.9712 g/cm^3 (dividing 12.35 by 3.12, then dividing 12.39 by 3.12). This is a variation of only 0.0129 g/cm^3.

As can be seen in the above comparison, the uncertainty in volume causes an uncertainty in the final result that is 4 times as great as the uncertainty in the same calculation caused by the uncertainty in mass. This is because there are four significant figures in the mass, and only three significant figures in the volume. The *actual* uncertainty incorporates both figures and could be estimated by comparing the (low)/(high) and the (high)/(low) figures. In the above example, these would be 12.35/3.14 (3.9395 g/cm^3) and 12.39/3.10 (3.9968 g/cm^3), or a total uncertainty of 0.0573 g/cm^3. Note the closeness of this value to that produced by the uncertainty in volume alone (in (a) above)! The correct value for the answer to this calculation would be 3.96 g/cm^3 (three significant figures). **The answer will contain only 3 significant figures, since the volume contained only 3 significant figures.**

EXAMPLE 3

A boy weighs his pet hamster on a laboratory balance and finds that it weighs 125.23 grams. He then weighs a sample of hamster food on an analytical balance and finds that it weighs 0.1273 grams. The hamster immediately eats the food, with no leftovers. What is the new weight of the hamster?

Answer: The 3 in 125.23 is *estimated*. It could be a 2 or a 4 or possibly a 1 or a 5. For this reason, it does not do any good to know the weight of the food to any *greater* accuracy than the hundredths place when adding it, since there will be uncertainty here anyway. Thus, *round off* the 0.1273 to 0.13 (125.23 + 0.13 = 125.36 g). The uncertainty is still in the hundredths place.

When adding or subtracting numbers, the answer is significant only to the fewest number of decimal places contained in any of the numbers being added or subtracted.

A second and equally important consideration of recording and using numbers is that virtually *all* numbers encountered in the laboratory represent quantities associated with *a specific material or phenomenon*. This means that value of the number is inseparably tied to the dimensions (frequently called "units") in which the quantity is measured. Suppose you were given the mass (or weight) of an object as 12, and asked how many of these you could carry. It makes a great difference whether the quantity is 12 milligrams, 12 grams, 12 ounces, 12 pounds, 12 kilograms, or 12 tons! **MEASURED NUMBERS ARE MEANINGLESS UNLESS THE PROPER DIMENSION (UNIT) ACCOMPANIES THE NUMBER. THESE DIMENSIONS (UNITS) ARE RETAINED IN CALCULATIONS.**

EXAMPLE 4

Seven corks weigh 28.23 grams. What is the average weight of a single cork?

Answer: (28.23 grams)/(7 corks) = 4.033 g/cork. Note that the 28.23 is a measured number and the 7 is a pure or exact number, so that there will be 4 significant figures in the answer. Many students find it helpful to verbalize the slash in the units as the words "for every." Thus, 4.033 g/cork would be read as 4.033 grams *for every* cork.

PROCEDURE

Numbers in parentheses refer to entry numbers on the data sheet.

This lab experience is designed to allow the student to become familiar with several common measuring devices which will be used throughout the course, and to develop an understanding of the inherent uncertainty in measuring with each. You will go into the lab and take measurements at stations which have been set up by your instructor before the lab. You will use the data for some simple calculations with a calculator. The class will then, as a whole, compare data to see where the numbers agree and where they disagree.

 I. GRADUATED CYLINDERS are glassware designed to transfer quickly a fairly accurately measured volume of liquid into another container and should not be used where *highly* accurate measurements are required.

Read the level of the meniscus (dividing line between water and air—your instructor will demonstrate the proper technique) for each graduated cylinder (1 and 2). Notice that the meniscus is curved at the middle (Fig. 1-2). The lowest point on this curve, not the upper edge, is always read as the volume. Avoid errors due to parallax; different readings are obtained if the eye's line of sight is not perpendicular to the scale.

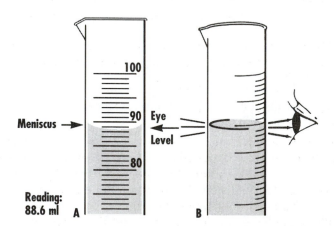

Figure 1-2. How to read a graduated cylinder.

II. BURETS are glassware designed to deliver conveniently a *very accurately* measured quantity of liquid into another container. The volume delivered is equal to the difference in the final and initial level readings. Read the liquid level in each of the three burets (3, 4, and 5) to the nearest 0.01 mL.

 III. THERMOMETERS are devices used to measure temperature. A *thermometer* will be used to measure the temperature of boiling water that is heated by a laboratory burner. Examine the burner and locate the gas and air valves (Fig. 1-3). Operate each valve before attaching the burner to the gas outlet. Close off the air and gas valves, attach hose to gas outlet on the burner and the desk, and open the desk valve about three-fourths of the way.

Strike a match and, while holding the lighted match to the side and just below the top of the barrel of the burner, gradually open the gas valve on the burner until a flame about 3 or 4 inches high is obtained. Now gradually open the air valve until a blue flame with an inner cone is obtained. Holding a piece of copper wire or a paper clip in the flame with tongs, locate the hottest and coolest parts of the flame by the degree of light shown by the copper. Is the hottest part of the flame at the top of the burner, in the middle of the blue flame, or at the top of the blue flame (6)?

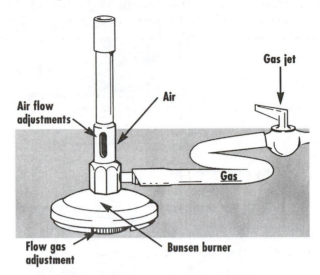

Figure 1-3. Bunsen burner features.

Set up a beaker on a wire gauze on an iron ring (Fig. 1-4). Fill the beaker about half full with water. Adjust your burner to give medium temperature and begin heating with water. Time can be saved if the water is heated while other parts of the experiment are being done. Periodically determine the temperature of the water with the thermometer. Be sure the mercury bulb is in the water, and be careful not to touch the walls of the beaker with the thermometer bulb. Record the boiling point of the water (7).

IV. BEAKERS are glassware designed for handling liquids (heating, reactions, etc.). The graduations are only *approximate*, and beakers are not designed as accurate measuring devices. Estimate the level of water in each beaker (8 and 9).

V. METER STICKS measure distance. Record the length of the object *in mm* (10).

VI. BALANCES are used in the laboratory for determining mass. Your instructor will demonstrate the use of the balance you are to use. You will determine the mass of an object with the balance, then determine the volume of the object by displacement, using a graduated cylinder. This mass divided by the volume gives its density, which is a physical characteristic of the substance. Weigh the object (11).

Fill a 100-mL graduated cylinder about half full of water and record the volume (12). Then carefully tilt the cylinder and slide the metal object down the side (to minimize splashing) and record the new liquid level (13). The metal must be completely submerged in the water. The difference in readings is the volume of the metal object (14).

Use a calculator to calculate the density of the object (15). Density = mass/volume. Also use a calculator to perform the mathematical operations listed at the bottom of the data sheet.

As your instructor indicates, read aloud the number you have recorded for each station, **exactly as you have recorded it.** Tables will be made for each station (one showing the complete calculator reading and the other showing the correctly rounded values). Note the range of measured values for each station. Comment on the "correct" value from these various tables (16). Compare the tables of calculated and rounded values. How do the correctly rounded values correlate with the total calculator readings (17)?

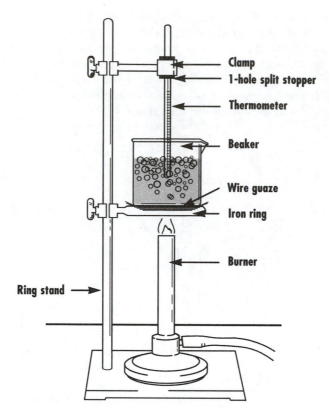

Figure 1-4. Set-up for thermometer bath.

PRE-LAB QUESTIONS

1. Under what circumstances is hair spray a laboratory hazard, as described in the **SAFETY CONSIDERATIONS** section?

    _____

    _____

    _____

    _____

    _____

    _____

2. Two students, working together, are asked to determine the average mass of a single cork in a bag of corks. One student determines the number of corks in the bag and the other student weighs the corks, so that the weight of all the corks can be divided by the number of corks. Comment on the *fundamental difference* in the nature of the numbers that each student obtained

    _____

    _____

    _____

    _____

    _____

    _____

3. Arrange the following in **increasing order** of the accuracy of the measurements which may be made with their graduation marks.

    buret, beaker, graduated cylinder

    _____ < _____ < _____

4. In only two parts of this laboratory procedure will the student actually perform operations *other than* reading scales that have been previously setup. What quantities are being measured in each of these instances?

    (1) _____

    (2) _____

Experiment 1
Data Sheet

Name _____

Date _____

INTRODUCTION TO LABORATORY MEASUREMENTS

graduated cylinders

(1) _____ mL

(2) _____ mL

beakers

(8) _____ mL

(9) _____ mL

burets

(3) _____ mL

(4) _____ mL

(5) _____ mL

length of object

(10) _____ mm

mass

(11) _____ g

thermometers

(6) _____

(7) _____ °C

(12) Initial cylinder level   _____ mL

(13) Final cylinder level   _____ mL

(14) Volume (subtract 12 from 13)   _____ mL

(15) Density (divide 11 by 14)   _____ g/mL (g/cm^3)

Calculations

Record both the calculator reading and the correctly rounded value.

a)   Volume of 400-mL beaker + volume of 100-mL cylinder (8 + 9)

calculator number _____ rounded number _____

b)   Volume of buret (3) × length of object (10)

calculator number _____ rounded number _____

c)   mass (11) ÷ length of object (10)

calculator number _____ rounded number _____

d)   Volume of buret (5) × Volume of buret (3)

calculator number _____ rounded number _____

e)   Density of metal object (15)

calculator number _____ rounded number _____

When each member of the class has performed the above calculations, the results will be collected into tables for examination.

(16) _____

_____

_____

(17) _____

_____

_____

POST-LAB QUESTIONS

1. Determine the correct answers to the calculation shown below *to the correct number of significant figures.*

    (a) $(32.453)(34.23 + 1.17)/(2.26 - 1.811)(0.00378) = $ _____

    (b) $(3.23 + 0.0018) = $ _____

    (c) $(82.97 - 0.012)/(1.100)(21.9) = $ _____

2. How many significant figures are in each of the following measurements, which have been taken with the indicated device?

    (a) 430 mL (beaker) _____

    (b) 237.35 g (triple beam balance) _____

    (c) 34.20 mL (buret) _____

Identification of a Substance

2

OBJECTIVE

To become familiar with the role of specific properties in the identification of substances.

APPARATUS AND CHEMICALS

balance
small test tubes (6)
spatula
dropper
25-mL graduated cylinder
125-mL conical flask and stopper
10-mL volumetric pipet
large test tubes (2)
ring stand
buret clamp
110°C thermometer
stopper with hole to hold thermometer
cork stoppers to fit small test tubes

rubber bulb for pipet
600-mL beaker
iron ring
wire gauze
burner and hose
right-angle bend with drain hose
boiling chips
test tube rack
one gram naphthalene
15 mL toluene
15 mL ethyl alcohol
5 mL xylene
unknown substance

SAFETY CONSIDERATIONS

 Be sure to wear eye protection to protect against possible glass breakage and splattering of chemicals.

 Do not pipet by using your mouth to draw the liquid into the pipet; rather, use a rubber bulb.

 When using the burner in the laboratory, be absolutely sure that loose hair and clothing and other flammable materials do not come close to the flame. Be sure the gas is turned off when you are through with the burner. Toluene, ethyl alcohol, and most of the liquid unknowns are flammable. Do not have these liquids near an open flame.

Do not throw lit or recently lit matches into any area or container where these flammable liquids have been discarded. Matches, paper, or other solid objects should *never* be discarded into sinks *under any circumstances*.

When melting the ends of glass tubes to seal the tubes for boiling point determinations, be aware that the glass **STAYS VERY HOT FOR SEVERAL MINUTES, even though it looks quite harmless.** *Very severe burns can result from touching this hot glass.*

 Protect your hands with a heavy towel when inserting the thermometer into the split rubber stopper or cork. Be sure to grasp the pipet near its top when you insert the pipet into the rubber bulb. If you grasp the pipet lower down, the act of attaching the rubber bulb can snap the pipet and drive the broken piece into your hand.

FACTS TO KNOW

The characteristics of a substance which enable us to distinguish it from other substances are known as properties. Table 2-1 gives a list of physical properties, subdivided into "specific properties" and "accidental properties."

Chemists often refer to "specific properties" as "intensive properties." These properties have numerical values that are independent of the quantity of the material that is present. "Accidental properties," whose numerical values vary depending on the quantity of material present in the sample, are frequently called "extensive properties."

Specific properties are of great value in the identification of a substance, for they can be measured objectively with scientific instruments, and they remain constant for a pure substance under fixed conditions of temperature and pressure. The accidental properties are of lesser value in the identification of a substance because accidental properties can vary from sample to sample of the same substance.

TABLE 2-1 PHYSICAL PROPERTIES

SPECIFIC PROPERTIES	ACCIDENTAL PROPERTIES
Solubility	Volume
Melting point	Weight
Boiling point	Shape
Density	
Refractive index	
Electrical conductivity	

The solubility of a substance in a pure solvent at a given temperature is the weight of that substance that will dissolve in a given weight (usually 100 g) of that solvent. Solubility is a specific property to be used in the identification of your unknown.

The density of a substance is defined as mass per unit volume:

$$D = \frac{M}{V}$$

Precise measurement of the density of a pure substance will aid in the identification of the substance.

The temperature at which a liquid and its solid are in equilibrium is known as the melting point of the solid and the freezing point of the liquid. The melting point of a solid is a specific property that can be conveniently measured and used in the identification of that substance.

When bubbles of vapor form within a liquid, rise freely to the surface, and burst, the liquid is said to boil. A liquid exposed to air will boil when its vapor pressure becomes equal to the pressure of the atmosphere. The normal boiling point of a liquid is that temperature at which the vapor pressure becomes exactly equal to the standard atmospheric pressure of 760 mm of mercury (one atmosphere). A liquid will boil at higher temperatures than normal under external pressure greater than one atmosphere; conversely, the boiling point of a liquid will be lower than normal when the external pressure is below one atmosphere.

The reference liquids you will be using in this experiment are common solvents in many everyday products. Water, the most common liquid on earth, is an excellent general solvent. Ethyl alcohol is a solvent found in some paints and paint removers, many liquid pharmaceuticals, and some cleaning fluids. Toluene (sometimes called toluol in industry) is a solvent found in glues, paints and paint removers, and spot removers. The reference solid, naphthalene, is recognizable by its characteristic odor as the material used in one type of mothball. Xylene is another organic solvent (like ethyl alcohol and toluene) that is a common solvent for paints and thinners.

PROCEDURE

Numbers in parentheses refer to entry numbers on the data sheet.

I. SOLUBILITY   Compare the solubility of naphthalene in three solvents: water, toluene, and ethyl alcohol. Do this by adding a few crystals of the solid to about 2 mL or 3 mL of each of the solvents contained in clean, *dry* test tubes. Try to use the same amount of naphthalene and solvent in each trial. Cork the test tube and shake briefly. Cloudiness indicates insolubility. Record your conclusions on the data page (1). Into three clean, *dry* test tubes place 2 mL or 3 mL of these same solvents, a different solvent in each test tube. Repeat the solubility tests, this time using two drops of xylene instead of naphthalene, and record your observations (2). Determine the solubility of your unknown in the manner described for naphthalene and xylene, and record (3).

II. DENSITY   Determine the density of your unknown substance obtained from the stock room for part I, using the directions under IIA if the unknown is a solid and those under IIB if the unknown is a liquid.

A. THE DENSITY OF A SOLID   Weigh to the nearest 0.01 g on the balance about 8 g of the unknown solid (4). If the unknown is unusually bulky, a smaller amount may be used. Half fill a 25-mL graduated cylinder with a liquid in which the solid was found to be insoluble in part I, and read the volume to the nearest 0.1 mL (5). Add the weighed solid to the liquid, being careful not to lose any material. Be sure that all of the solid is beneath the surface of the liquid and the air bubbles have all cleared the surface of the liquid. Carefully thump the graduated cylinder to dislodge any air bubbles trapped in the solid. Read the new volume to the nearest 0.1 mL (Fig. 2-1). The difference in the two volumes represents the volume of the solid (7). Calculate the density of the solid in g per mL (8).

If air bubbles were trapped in the solid beneath the liquid, would the volume measurement be larger, smaller, or unchanged (9)? What would be the effect of the error on the calculated density: to make it larger, smaller, or unchanged (9)?

B. THE DENSITY OF A LIQUID   Carefully weigh a clean, dry conical flask and its stopper to the nearest 0.01 g on the balance (10). Pour at least 15 mL of the unknown liquid into a clean, dry test tube. Hold the pipet near its top with one hand and with the other hand, barely place the rubber bulb onto the pipet. Squeeze the rubber bulb slightly, and insert the constricted end of the clean and dry pipet to the bottom of the liquid in the test tube. Slowly release the pressure on the bulb and carefully draw the liquid into the pipet (making sure the tip of the pipet stays beneath the surface of the liquid) until the liquid level is about 2 cm above the etched line. Push the rubber bulb off with a finger. In the same motion, slide the finger over the top of the pipet and clamp down gently. Raise the pipet tip to a position above the surface of the liquid. Hold the pipet vertically and rub your finger slowly across the top of the pipet while releasing the pressure of your fingertip *slightly*. Allow the level of the liquid to fall until the bottom of the meniscus lies exactly on the etched line, as shown in Figure 2-2. Wipe off any drops adhering to the tip of the pipet and hold the pipet in a vertical position over the weighed flask. Remove your finger from the upper end and allow the liquid to drain into the flask. Allow the pipet to drain for 10 seconds after the flow of the liquid has stopped, and then touch the tip to the side of the flask to remove any drops adhering to the tip. Do *not* try to force the last of the liquid from the tip; the pipet is designed to deliver exactly 10 mL of liquid by the action of gravity alone. Weigh the flask and its contents (11). What is the weight of the liquid (12)? From the data obtained, calculate the density of the unknown liquid by dividing its weight by its volume (10 mL) (13).

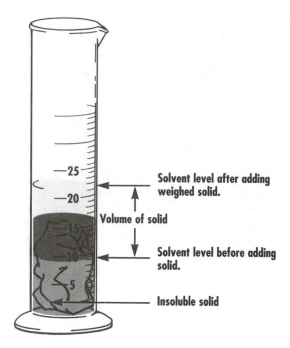

Figure 2-1. Set-up for the measurement of the density of a solid.

III. MELTING POINT (for Solid Unknowns) If your capillary tubes are open on both ends, gently heat one end of the capillary tube in the edge of the flame of a Bunsen burner until the end completely closes. Rotate the tube between your thumb and fore finger during heating to prevent it from bending. Keep the end to be heated higher than the end you are holding.

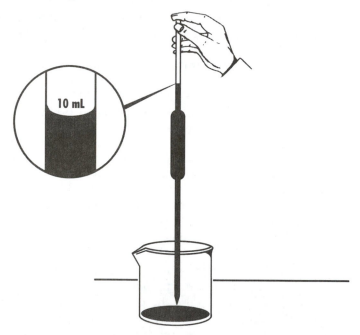

Figure 2-2. Proper reading of a pipet.

Pulverize about 1 g of the sample with a clean mortar and pestle. Carefully push the open end of the capillary tube into the powdered sample, forcing a portion of the sample into the tube. Invert the tube and flick the tube with your finger, or rasp the top with a file to vibrate the sample into the closed end. Repeat until the tube contains sample to a depth of about $1/2$ cm. Do not press too much sample into the capillary tube during each step. It will pack, and then it cannot be forced to the bottom of the tube.

Place a small rubber band 3 cm to 5 cm above the bulb of the lower end of a thermometer. Insert the capillary tube under the rubber band with the closed end and sample near the bulb. Place the thermometer into a split cork or rubber stopper so that it may be secured in a buret clamp. Place the thermometer and capillary tube in a 250-mL beaker containing enough water to cover the bulb and portion of capillary containing the sample. Be sure that the open end of the capillary is above the water level. Figure 2-3 shows the arrangement of equipment for melting point determination.

Heat the water slowly; try for a rate of increase of about 1° per 20 seconds. Observe the solid in the capillary very carefully. The melting point is the temperature at which the solid drops (or sinks) in the capillary tube. Note the temperature at that moment. This is the melting point of the solid (14).

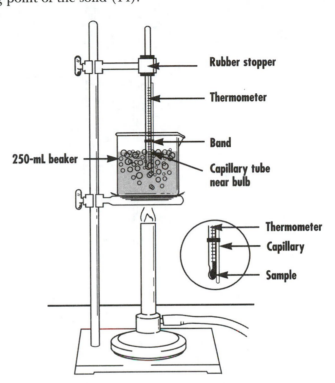

Figure 2-3. Set-up for determination of a melting point.

IV (a). BOILING POINT (for Liquid Unknowns) Add unknown liquid to a test tube to depth of about 3 cm. The test tube is fitted with a two-hole stopper, a thermometer, and a right-angle glass tube, as shown in Figure 2.4. Add a very small piece of boiling chip to the liquid to insure even boiling. Adjust the thermometer so that the bulb is 1 cm above the surface of the liquid. Clamp the test tube to the ring stand. Connect the length of rubber tubing to the glass tube and put the end of the tubing in the sink. Support a 600-mL beaker of water on a ring and wire gauze and immerse the bottom fourth of the test tube in the water. The test tube should not touch the beaker. Heat the water gradually and watch for changes. Record the temperature at which the liquid in the test tube boils freely (15). List some likely sources of error involved in this determination (16).

(b). MICROBOILING POINT A second and more sensitive method of determining a boiling point which requires much less unknown liquid (and hence reduces the quantity of vapor produced) involves the use of a very narrow glass tube (15 cm × 5 mm i.d.) filled about one third full with the unknown liquid. Such a tube can be easily made by sealing one end of a piece of glass tubing in a very hot Bunsen burner. [**CAUTION: THE GLASS REMAINS VERY HOT FOR SEVERAL MINUTES AFTER THE TUBE HAS BEEN SEALED!**] This is attached to a thermometer as shown in Figure 2-5. A sealed capillary tube is place in the liquid *open end down*, and the test tube and thermometer is immersed in a beaker of high boiling non-flammable liquid (such as mineral oil). As the mineral oil is heated and the air in the capillary tube warms up, bubbles of air slowly are forced out of the mouth of the capillary tube. When the temperature of the mineral oil rises to the boiling point of the liquid in the test tube, bubbles will emanate from the capillary tube *very rapidly* (it will look like an air hose has been inserted into the test tube). When this occurs, immediately remove the heat from the mineral oil bath and allow it to cool, as you continually stir the mineral oil bath. As the mineral oil cools, formation of bubbles will cease and the liquid will begin to rise in the capillary tube. The boiling point is recorded as the temperature on the thermometer when the level of the liquid in the capillary tube has risen to exactly the level of the remaining liquid in the glass tube (NOT the level of mineral oil in the beaker).

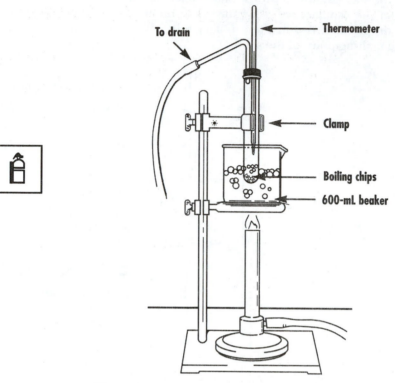

Figure 2-4. Set-up for determination of a boiling point.

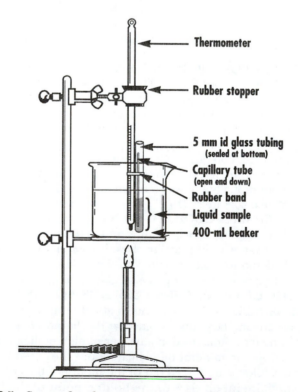

Figure 2-5. Set-up for determination of a microboiling point.

V. REPORT Your unknown is one of the substances in the following list. Compare the set of properties you have determined for your unknown with those listed and identify the unknown (17). Turn in the data sheet to your instructor with questions (1) through (9), (14), and (17) answered if your unknown is a solid, or (1) through (3), (10) through (13), (15), (16), and (17) answered if your unknown is a liquid.

TABLE 2-2 PHYSICAL PROPERTIES OF SELECTED PURE SUBSTANCES*

SUBSTANCE	DENSITY (g/mL)	MELTING POINT (°C)	BOILING POINT (°C)	SOLUBILITY		
				Water	Toluene	Alcohol
Acetone	0.79	−95	56	S	S	S
Benzophenone	1.15	48	306	I	S	S
Cyclohexane	0.78	6.5	81.4	I	S	S
p-Dibromobenzene	1.83	86.9	219	I	S	S
p-Dichlorobenzene	1.46	53	174	I	S	S
Diphenylamine	1.16	53	302	I	S	S
Diphenylmethane	1.00	27	265	I	S	S
Ether, ethyl propyl	1.37	−79	64	S	S	S
Hexane (component of gasoline)	0.66	−94	69	I	S	S
Isopropyl alcohol	0.79	−98	83	S	S	S
Lauric acid (found in coconut oil)	0.88	43	225	I	S	S
Magnesium nitrate·6H$_2$O	1.63	89	330**	S	I	S
Methyl (wood) alcohol	0.79	−98	65	S	S	S
Methylene chloride	1.34	−97	40.1	S	S	S
Naphthalene	1.15	80	218	I	S	SI
α-Naphthol	1.10	94	288	I	S	S
Phenyl benzoate	1.23	71	314	I	S	S
Propionaldehyde	0.81	−81	48.8	S	I	S
Stearic acid (found in animal fats)	0.85	70	291	I	S	SI
Thymol	0.97	52	232	SI	S	S
p-Toluidine	0.97	45	200	SI	S	S

* Symbols used in this table: S = soluble; SI = slightly soluble; I = insoluble

** Decomposes

PRE-LAB QUESTIONS

1. Comment on the usefulness of specific properties (intensive properties) compared to those of accidental properties (extensive properties) to an analyst.

    _____

    _____

    _____

    _____

    _____

    _____

2. Classify each of the properties as specific (S) or accidental (A).

    _____ boiling point   _____ color   _____ mass

    _____ volume   _____ density   _____ ignition temperature

3. The atmospheric pressure in Denver (the "Mile High City") is significantly lower than it is in New York City (very close to sea level). How would the boiling points of the same liquid compare in the two cities, and WHY?

    _____

    _____

    _____

    _____

    _____

    _____

4. In the microboiling point method of determining the boiling point of an unknown liquid, what temperature is recorded as the boiling point of the liquid?

    _____

    _____

    _____

    _____

    _____

    _____

Experiment 2
Data Sheet

Name _____

Date _____

IDENTIFICATION OF A SUBSTANCE

SOLUBILITY

	Water	Toluene	Alcohol
(1) Naphthalene	_____	_____	_____
(2) Xylene	_____	_____	_____
(3) Unknown	_____	_____	_____

DENSITY

(A) For solid unknowns:

(6) Final volume of liquid in cylinder _____ mL

(5) Initial volume of liquid in cylinder _____ mL

(7) Volume of liquid displaced by solid _____ mL

(4) Weight of unknown _____ g

(8) Density of solid _____ g/mL

(9) Error resulting from air bubbles trapped beneath liquid

Volume: _____ mL

Density: _____ g/mL

(B) For liquid unknowns:

(11) Weight of flask and unknown _____ g

(10) Weight of empty flask _____ g

(12) Weight of liquid _____ g

(13) Density of unknown _____ g/mL

Melting Point (Solids Only)

(14) Melting point _____ °C

Boiling Point (Liquids Only)

(15) Boiling point _____ °C

(16) Sources of error in boiling point determination

(17) Identification of unknown

POST-LAB QUESTIONS

1. Two accidental (extensive) properties are volume and mass. By themselves, neither can be used as an identifying characteristic of a material. How can these be combined to give a characteristic that *can* be used as an identifying property?

    _____

    _____

    _____

2. WHY does combining the quantities of mass and volume as described in question 1 result in a much more useful property?

    _____

    _____

    _____

3. In the melting point determination and the boiling point determination, the answer may vary depending on the care with which the measurement is carried out. What *procedural variations* might cause one student's measurement to differ from that of another student's value? [These are commonly referred to as "sources of error."]

    _____

    _____

    _____

The Separation of Components of Mixtures

OBJECTIVE

To become familiar with the techniques of filtration, extraction, and sublimation as means of separating the components of mixtures.

APPARATUS AND CHEMICALS

evaporating dish (2)
clay triangle (2)
25-mL graduated cylinder
glass stirring rod
tongs
watch glass

ring stand (2)
iron ring (2)
burner and hose
mixture of sodium chloride,
   ammonium chloride, and
   silicon dioxide

SAFETY CONSIDERATIONS

 The Bunsen burner is used in this experiment. Care should be exercised in igniting and using the burner. Don't forget to turn the gas off after use.

 Dust and the possibility of broken glassware require eye protection.

 Care should be taken to avoid touching the hot metal ringstand. See Figure 3-2.

 Ammonium chloride is a mild respiratory irritant. Breathing of the fumes should be avoided.

FACTS TO KNOW

When two or more substances that do not react chemically are blended, the result is a mixture in which each of the component substances retains its identity and its fundamental properties. The properties observed for the mixture are simply a combination of the properties of components.

If one of the components of a mixture is present in by far the largest amount, the mixture may be regarded as that substance in impure form and the components in small amounts are impurities in the larger component.

The separation of the components of mixtures, including the purification of impure substances, is a problem frequently encountered in chemistry. The basis of the separation of the components of a mixture is the fact that each component has a different set of fundamental properties. The components of mixtures are pure substances which are either elements or compounds. Each element or compound has fundamental properties by which it may be identified. Under the same conditions of temperature and pressure, the fundamental properties of every sample of the same pure substances are identical.

For example, every sample of sugar under normal conditions is a solid composed of transparent crystals of the same type. The crystals in confectioner's sugar are smaller than those in regular granulated sugar, but size of particles is not a fundamental property.

Every crystal of a pure substance will melt at a given temperature. At a given external pressure, every pure substance boils at a given temperature, provided that it does not decompose before it boils. At a given temperature, every substance has a particular solubility. This solubility can be expressed in grams of solute per 100 grams of solvent. The composition of each pure substance is constant.

Although these and other characteristics can be used to identify a particular substance, we will be concerned in the experiment with the separation of a mixture into its components, not with the identification of the substances. Techniques used to separate mixtures rely on differences in fundamental properties. These techniques include the following:

Distillation is the purification of a liquid by heating it to its boiling point, condensing it, and collecting the vapors. Separation of two or more liquids requires that they have different boiling temperatures. All boiling temperatures can be reduced by decreasing the pressure on the liquid.

Extraction is the removal of one substance from a mixture because of its greater solubility in a given solvent.

Filtration is the process of removing or "straining" a precipitate or suspended solid from a liquid.

Centrifugation is the process of separating a suspended solid from a liquid by whirling the mixture at a high speed.

Sublimation is the process of going directly from the solid state into the gaseous state, without melting (passing through the liquid state). Only a relatively few pure substances do this at normal pressures. Therefore, if one component of a mixture sublimes, this property may be used to separate it from the other components. Iodine, naphthalene, and ammonium chloride (NH_4Cl) are substances that sublime.

The mixture you will separate contains three components: sodium chloride, ammonium chloride, and silicon dioxide.

TABLE 3-1 PROPERTIES OF SODIUM CHLORIDE, AMMONIUM CHLORIDE, AND SILICON DIOXIDE

SPECIES	FORMULA	MOLECULAR WEIGHT	SOLUBILITY (g) IN 100 g WATER AT 25°C	MELTING POINT (C°)	APPEARANCE
Sodium chloride (salt)	NaCl	58.5	35	801	white crystals
Ammonium chloride (sal ammoniac)	NH_4Cl	53.5	37	sublimes 350	white crystals
Silicon dioxide (sand)	SiO_2	60.1	insoluble	1600	white crystals

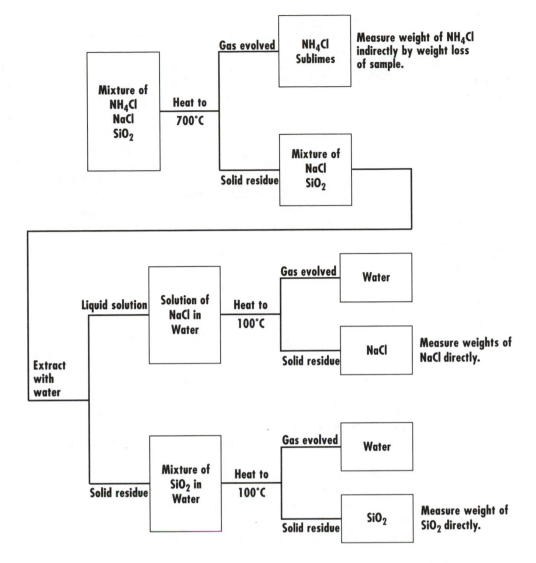

Figure 3-1. Scheme to separate ammonium chloride, sodium chloride, and silicon dioxide based on properties listed in Table 3-1.

PROCEDURE

Numbers in parentheses refer to entry numbers on the data sheet. Record all weights on the data sheet.

I. Carefully weigh a clean, dry evaporating dish to 0.01 g (1). Use this evaporating dish to obtain from the instructor a sample of the mixture of sodium chloride, silicon dioxide, and ammonium chloride. Weigh the evaporating dish and sample to 0.01 g and record (2). Subtract the weight of the evaporating dish from the weight of the evaporating dish and the sample to obtain the weight of the sample (3).

II. Place the evaporating dish and sample on a clay triangle, ring, and ring stand assembly *in the hood* (Fig. 3-2). Heat with a burner (slowly at first) until white fumes cease (about 15 minutes). You may need to remove the flame after about 10 minutes of heating to gently stir the mixture; then apply the heat again. Heat until no more white powder condenses on the stirring rod.

Let the evaporating dish cool until you can hold it firmly in your hand. Do not use tongs to lift the evaporating dish. Weigh the evaporating dish and solid and record (4). Subtract answer number (4) from answer number (2) to obtain the weight of ammonium chloride, NH₄Cl, in the sample (5).

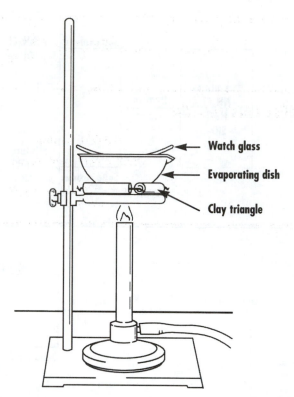

Watch glass

Evaporating dish

Clay triangle

Figure 3-2. Apparatus for heating samples. (a) To sublime NH$_4$Cl, remove watch glass, place apparatus in hood, and heat strongly. (b) To evaporate water from SiO$_2$ or NaCl, heat gently; a watch glass can be used to contain splattering near the end.

III. Weigh the other evaporating dish and record its weight (6). Add 30 mL of water to the solid in the original evaporating dish. Stir gently for 5 minutes. Pour the liquid only (decant) into the second preweighed evaporating dish and leave the solid in the original evaporating dish. Add 10 more mL of water to the solid in the evaporating dish, stir for at least 5 minutes, and pour the liquid into the second evaporating dish, as before. Repeat with another 10 mL of water. This process extracts the sodium chloride from the sand. The objective now is to recover these two substances so they can be weighed.

IV. Set the evaporating dish containing the solution of sodium chloride and water on a clay triangle on a ring and ring stand assembly. Evaporate the water. Reduce the heat at the beginning to avoid boiling over and reduce it again later to avoid splattering near the end of the drying. While you are evaporating the water from the NaCl solution, you may proceed to step V. Near the end, a watch glass may be placed over the evaporating dish to prevent splattering. The watch glass also indicates when the salt is dry because water will cease to condense on the watch glass. When the salt is dry, remove the heat and let the evaporating dish cool until you can hold it tightly in your hand. Weigh and record (7). Subtract the weight of this evaporating dish (6) from (7) to obtain the weight of sodium chloride in the mixture (8).

V. Place the evaporating dish containing the solid (SiO$_2$) on a clay triangle and heat *slowly* with stirring until the lumps break up and the sand appears to be dry. Then heat the bottom of the evaporating dish to redness and maintain this heat for 10 minutes. Take care not to break the evaporating dish. When the solid is dry, remove the heat and let it cool. Weigh and record (9). Subtract the weight of the original evaporating dish (1) from (9) to obtain the weight of the silicon dioxide (10).

VI. Although three independent determinations, as have been done in this experiment, are not necessary, it is generally considered better technique than to determine two quantities and subtract to obtain the third because if you miss one percentage, you necessarily miss two percentages when the sum of the two percentages is subtracted from 100% to obtain the third percentage.

Calculate the percentage of each substance in the mixture by using the following formula:

$$\% \text{ component} = \frac{\text{grams of component} \times 100\%}{\text{grams of sample}}$$

EXAMPLE CALCULATION

10.63 g sample

2.63 g NaCl recovered

$$\% \text{ NaCl} = \frac{2.63 \text{ g}}{10.63 \text{ g}} \times 100\% = 24.7\%$$

PRE-LAB QUESTIONS

1. Which of the three components of the mixture is to be separated from the others by virtue of its *solubility* in a particular solvent, and what is that solvent?

   _____

   _____

2. What unique property of ammonium chloride (NH_4Cl) allows it to be separated from the other components of the mixture?

   _____

   _____

3. How do properties of a mixture differ from the properties of the individual pure substances that make up the mixture?

   _____

   _____

   _____

   _____

4. How is the mass of SiO_2 contained in a sample of the mixture determined?

   _____

   _____

   _____

   _____

Experiment 3
Data Sheet

Name _____

Date _____

THE SEPARATION OF COMPONENTS OF MIXTURES

(2) Weight of evaporating dish #1 and sample   _____ g

(1) Weight of evaporating dish #1   _____ g

(3) Weight of sample (subtract 1 from 2) g

(2) Weight of evaporating dish #1 and sample (NH_4Cl, NaCl, SiO_2)   _____ g

(4) Weight of evaporating dish #1 and sample (NaCl, SiO_2)   _____ g

(5) Weight of NH_4Cl (subtract 4 from 2)   _____ g

(7) Weight of evaporating dish #2 and residue (NaCl)   _____ g

(6) Weight of evaporating dish #2   _____ g

(8) Weight of residue (NaCl) (subtract 6 from 7)   _____ g

(9) Weight of evaporating dish #1 and residue (SiO_2)   _____ g

(1) Weight of evaporating dish #1   _____ g

(10) Weight of residue (SiO_2) (subtract 1 from 9)   _____ g

— (11) Percentage NH_4Cl _____

— (12) Percentage NaCl   _____

— (13) Percentage SiO_2 _____

(14) Total percentage recovered   _____

Do the calculations in an organized fashion below:

10 gm

POST-LAB QUESTIONS

1. A sample of a mixture is found to consist of 1.24 g of sugar, 2.87 g of salt, and 3.12 g of baking soda. What is the per cent composition of this mixture?

   % salt = _____

   % sugar = _____

   % baking soda = _____

2. It is not always possible to *see* all of the white fumes escaping when the ammonium chloride sublimes, nor does it always show up on the stirring rod. What steps could one take in order to be reasonably sure that *all* of the NH$_4$Cl had been sublimed away in step II of the procedure?

   _____

   _____

   _____

3. Why do you think the NH$_4$Cl was separated from the mixture *before* the NaCl was separated. Why would it most likely not work as well to try and separate the NaCl first?

   _____

   _____

   _____

   _____

4

Types of Chemical Reactions

OBJECTIVE

To conduct a series of simple experiments, make observations, write balanced equations for the reactions, and classify the reactions as one or more of the types described below.

APPARATUS AND CHEMICALS

test tubes
wood splint
test tube clamp
Bunsen burner
matches
magnesium ribbon
zinc metal
red litmus paper
hydrochloric acid (6 M)
sodium oxide (solid)
manganese dioxide (solid)

hydrogen peroxide (3%)
iron wire
copper sulfate solution (1 M)
sulfuric acid (1 M)
lead nitrate solution (0.1 M)
calcium carbonate (solid)
hydrated copper(II) sulfate (solid)
sodium carbonate solution (0.1 M)
ferrous ammonium sulfate solution (0.1 M)
potassium permanganate solution (0.1 M)

SAFETY CONSIDERATIONS

 When using a burner or any open flame in the laboratory, be absolutely sure that loose hair and clothing do not come close to the flame. It is easy to forget this simple rule when concentrating on measurements.

 Be sure to wear eye protection. Avoid looking directly at the magnesium ribbon while burning it. Burning magnesium emits brilliant light of sufficient intensity to cause temporary impairment of eyesight if viewed directly.

 Objects such as test tubes can be very hot even when they do not look hot. Handle all objects with tongs or test tube holders.

Be sure the gas is turned off when you are through with it because in a closed laboratory a small leak can create an explosive mixture.

 Gloves and a laboratory coat will provide additional safety from any corrosive chemical that may splatter on you. Sulfuric acid will burn and dehydrate the skin. Wash the acid off of your skin immediately with much running water. Sulfuric acid will eat through your clothes. Hydrochloric acid can also burn the skin and destroy your clothes.

 Potassium permanganate is a strong oxidizing agent. It will permanently bleach any clothing it contacts and will cause holes if not washed off clothing quickly.

FACTS TO KNOW

The elements and compounds may be characterized by the ways in which they react. Most inorganic substances can be classified by five common types of reactions. Those reactions are:

1) **Addition Reactions**
   Reactions where two or more substances combine to give one substance. Note the examples below:

   (1) $S + O_2 \rightarrow SO_2$

   (2) $SO_3 + H_2O \rightarrow H_2SO_4$

2) **Decomposition Reactions**
   A reaction in which one substance breaks down into two or more substances, as the following examples indicate:

   (3) $CaCO_3 \rightarrow CaO + CO_2$

   (4) $2\ NaHCO_3 \rightarrow Na_2CO_3 + H_2O + CO_2$

3) **Oxidation-Reduction Reactions**
   Reactions in which substances experience a change in their oxidation state are oxidation-reduction reactions. Both oxidation and reduction must occur simultaneously. Examples include:

   (5) $2\ Na + Cl_2 \rightarrow 2\ NaCl$

   (6) $Zn + 2\ HCl \rightarrow ZnCl_2 + H_2$

4) **Metathetical (double displacement) Reactions**
   In these reactions, the ions which make up one compound exchange partners with the ions of another compound. The formation of a precipitate (solid) and the evolution of a gas are two signs that a metathetical reaction has occurred. Examples include:

   (7) $AgNO_3 + KBr \rightarrow AgBr$ (a solid) $+ KNO_3$

   (8) $K_2S + 2\ HCl \rightarrow H_2S$ (a gas) $+ 2\ KCl$

5) **Acid-Base Reactions (neutralization)**
   A simple acid-base reaction involves the transfer of a hydrogen ion from the acid to the base. As a result, the acidity of the solution has undergone a change. Examples include:

        acids   bases

   (9) $HCl + NaOH \rightarrow NaCl + H_2O$

   (10) $H_2SO_4 + Ba(OH)_2 \rightarrow BaSO_4 + 2\ H_2O$

Some reactions may be classified by more than one type. For example, reaction 1 above has been classified as an addition reaction. It is also an example of oxidation-reduction. Sulfur and oxygen are both in the zero oxidation state prior to the reaction. After sulfur and oxygen reacted, the sulfur exhibits a $+4$ oxidation state and the oxygen a -2 oxidation state. Also reaction number 6 above is sometimes referred to as a single displacement reaction.

There are a number of observations which give evidence that a chemical reaction has occurred.

1) A change in the acidity of solutions. Litmus paper can be used to detect any change in the acidity of solutions before and after mixing. Acids turn blue litmus paper pink, and bases turn pink litmus blue. Students should test solutions prior to mixing by noticing their effect on litmus. Similarly, test the solution with litmus paper after mixing.

2) Formation of a precipitate (solid). If a precipitate forms when two solutions are mixed, a reaction has occurred.

3) Evolution of a gas. The formation of a gas is evidence that a reaction has occurred. The evolution of a gas can be detected by its color, odor, acidity, and/or by its combustibility.

4) Color change. Brightly colored solids often will undergo a change in color upon heating. This color change indicates that the solid has undergone decomposition.

5) Observable temperature change. All reactions are accompanied by energy changes of some sort. These may include sound (ex. an explosion), light (a fire), electric current (battery discharging), and heat. Many reactions give off a combination of forms of energy (an explosion produces light, heat, and sound). Sometimes the only observable change accompanying a reaction is the absorption or release of energy. For example, when a solution of aqueous HCl is added to a solution of aqueous NaOH to form an aqueous solution of NaCl (and a little more water), it *looks like* water being added to water making more water, except that it gets hot!

PROCEDURE

Below are a series of reactions to be conducted. Follow the directions for each reaction and use the data sheet to provide the following:

(a) your observations   (b) the reaction type   (c) a balanced equation

(1) Demonstration by instructor: Secure a piece of magnesium ribbon about 3 inches long, hold one end with tongs, and ignite one end of it with a match. *CAUTION:* Look at this reaction out of the side of your eyes rather than looking directly at the hot spot in the center of the flame.

(2) Place a few small pieces of zinc metal in a clean test tube and add a few drops of dilute hydrochloric acid (HCl).

(3) Add 2 mL of 3% H_2O_2 to a small test tube. Use a spatula to add a small amount of MnO_2 (0.1 g). Hold a glowing splint in the end of the test tube to test whether the gas given off supports combustion. MnO_2 is a catalyst in this reaction. Explain this in the observations section.

(4) Place a few pieces of iron wire in a test tube containing about 5 mL of copper sulfate ($CuSO_4$) solution. Allow several minutes for a reaction to occur.

(5) Add a few drops of 1 M sulfuric acid (H_2SO_4) solution to a test tube containing about 2 mL of lead nitrate ($Pb(NO_3)_2$) solution.

(6) Place a small amount of calcium carbonate ($CaCO_3$) solid in a clean, dry test tube and slowly add dilute hydrochloric acid dropwise.

(7) Place about 1 g of solid copper sulfate hydrate ($CuSO_4 \cdot 5H_2O$) in a clean, dry test tube and heat for several minutes.

(8) Test the acidity of tap water using red litmus paper. Dissolve about 0.1 g of sodium oxide (Na_2O) solid in the water and repeat the test with red litmus paper.

(9) Place about 3 mL of a solution of ferrous ammonium sulfate ($Fe(NH_4)_2(SO_4)_2$) in a test tube. Add dropwise a solution of potassium permanganate ($KMnO_4$).

(10) Place about 3 mL of a solution of Na_2CO_3 in a clean test tube and test its acidity with a piece of red litmus. Add slowly about 20 drops of dilute hydrochloric acid and test the acidity of this solution with red litmus paper.

PRE-LAB QUESTIONS

1. What precaution must be taken when burning magnesium ribbon is observed?

    _____

2. State briefly what should be done if sulfuric acid is spilled on the skin.

    _____

3. List two observations which you can make which indicate that a chemical reaction has occurred.

    1. _____

    2. _____

4. State the color which litmus paper will change to in:

    2. acid solution _____

    3. basic solution _____

Experiment 4
Data Sheet

Name _____

Date _____

TYPES OF CHEMICAL REACTIONS

Observations	Reaction Type	Complete and Balance*
(1)		$2\ Mg + O_2 \rightarrow$
(2)		$Zn + 2\ HCl\ (aq) \rightarrow$
(3)		$2\ H_2O_2\ (aq) \overset{MnO_2}{\rightarrow}$
(4)		$Fe + CuSO_4\ (aq) \rightarrow$
(5)		$Pb(NO_3)_2\ (aq) + H_2SO_4\ (aq) \rightarrow$
(6)		$CaCO_3\ (s) + 2\ HCl\ (aq) \rightarrow$
(7)		$CuSO_4 \cdot 5H_2O_{(s)} \overset{\Delta}{\rightarrow}$
(8)		$Na_2O_{(s)} + H_2O \rightarrow$
(9)		$FeSO_4\ (aq) + KMnO_4\ (aq) \rightarrow$
(10)		$Na_2CO_3\ (aq) + 2\ HCl\ (aq) \rightarrow$

* An aqueous (water) solution is represented by the symbol (aq).

POST-LAB QUESTIONS

1. Manganese dioxide (MnO_2) was added to the test tube containing hydrogen peroxide (H_2O_2). Why was this added? _____

   What can the manganese dioxide be called when it is used for this purpose?

   _____

2. What color did the litmus paper turn to when you tested tap water? _____

   Make a guess as to why tap water gave this reaction. _____

   _____

3. Write the name and formula for the chemical substance which is always produced in a neutralization reaction.

   1. name _____

   2. formula _____

5 Molecular Shapes

OBJECTIVE

To learn how the shapes of some simple molecules may be determined, and to examine some of the factors that determine the shape of the molecules.

APPARATUS AND CHEMICALS

Styrofoam balls of various sizes, toothpicks.

SAFETY CONSIDERATIONS

Be careful with the toothpicks. The ends are sharp and can cause injury.

FACTS TO KNOW

Many of the properties of substances are determined in part by the shapes of the molecules. These include physical properties, such as boiling point, melting point, densities, etc. and chemical properties such as reactivities. It is often difficult to get a good feel for the actual shape of a molecule from the drawings one might make, or even from carefully produced artists depictions, as one sees in many textbooks, because drawings and textbook illustrations are limited to 2-dimensional representations of 3-dimensional entities. Three-dimensional space filling models of molecules are available and are often very helpful in understanding the relation between structure and properties, but these are often too expensive for many students to have for study purposes. This exercise will show how to construct meaningful models with cheap, readily available materials.

The procedures used in predicting the shapes of simple molecules for this experiment are summarized in the Valence Shell Electron Pair Repulsion theory (VSEPR Theory). This is described in many introductory textbooks and relies on the (very logical) observation that since electrons are all negatively charged, they will stay as far apart from one another as they can, given the requirements of bonding. The bonding requirements (and other electronic considerations in the atom) lead to the assumption that electrons will be in pairs in both bonds and in the outermost electron shell when they are not engaged in bonding. The heart of the VSEPR theory is the electron dot diagram of the molecule, which shows how many bonds are in a molecule, and which atoms are bonded together, as well as the electrons in the valence shell of the atom that are not involved in bonding (which are called "non-bonded pairs" or "lone pairs"). Techniques for drawing electron dot diagrams (called "Lewis diagrams" after the noted chemist Gilbert N. Lewis, who developed them) may be found in any introductory textbook and will not be discussed here. We will examine some of the simple structures that can be drawn, however.

A great many molecules consist of one central atom, which is bonded to several atoms which themselves are only bonded to the central atom. Atoms bonded only to the central atom are called terminal atoms, while atoms bonded to more than one other atom are called central atoms. The bonds consist of one or more electron pairs, and there are frequently also non-bonded pairs of electrons in the molecule as well. *Non-bonded electron pairs on central atoms play a very important role in determining the shape of a molecule!* Non-bonded electron pairs on terminal atoms play almost no role in affecting the structure of a molecule. Most central atoms in a molecule will have between two and six electron pairs in their valence shells. It is the total number of valence electron pairs, *as well as how many are non-bonded pairs and how many are bond pairs* that in large part determines the shape of the molecule. Shown below are some examples of molecules containing bonds and non-bonded electron pairs. In these structures, a bond consists of one electron pair. A double bond consists of 2 electron pairs shared between the same two atoms. *A double bond counts only as one pair of electrons insofar as shape is concerned—see formaldehyde below.* Reminder: Non-bonded pairs of electrons on terminal atoms do not affect molecular shape, but non-bonded pairs on the central atom are very important. The summary under each molecule represents the electron pairs *on the central atom.*

CH_4
(methane)
4 bond pairs
0 lone pairs

H_2O
(water)
2 bond pairs
2 lone pairs

NH_3
(ammonia)
3 bond pairs
1 lone pair

CH_2O
formaldehyde
3 bond pairs*
0 lone pairs

ClF_3
Chlorine trifluoride
3 bond pairs
2 lone pairs

CO_3^{2-}
carbonate ion
3 bond pairs*
0 lone pair

Both bond pairs and lone pairs of electrons occupy space around a central atom, and the lone pairs actually occupy *more* space than do the bond pairs. Thus, the pairs of electrons will adopt shapes that maximize the distance between one another. This lab is aimed at learning what these shapes are, for various numbers of electron pairs around a single central atom.

PROCEDURE

Obtain from your instructor a collection of various sized styrofoam balls and a number of toothpicks. In the following exercises, try and minimize the number of times a toothpick is inserted into a given ball, as the balls can deteriorate and be difficult to work with after a while.

1. Stick 4 toothpicks halves into a 1½ inch sphere, all at 90° to one another and in the same plane (this is a square planar arrangement). Use the 90° angle template provided. (TO USE THE TEMPLATE, YOU MUST ENVISION THE TOOTHPICKS COMING TOGETHER AT THE CENTER OF THE SPHERE. SEE THE DIAGRAM ON THE NEXT PAGE.)

   Place a small sphere on each toothpick piece so that each small sphere touches the large sphere (you may have to shorten the toothpicks). This is a square planar shaped particle. Sketch your model on the answer sheet.

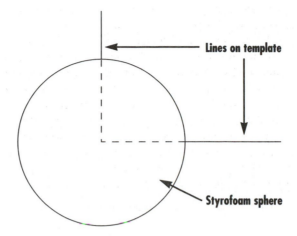

Replace 1 small sphere with a 2 inch sphere. Does it produce different shapes if you choose to replace a different small sphere, as long as only one is replaced? If you can get the same shape you had before by simply *rotating* the shape you have, it is not a uniquely new shape. Sketch your model(s).

Replace a second small sphere with a larger one. Can you make more than one different shape by replacing exactly 2 spheres? How many unique shapes can you make by replacing 2 small spheres with larger spheres? Again, if you can get a shape you already have simply by rotating or turning the model it is not a unique shape. Sketch your model(s).

Repeat the process above, but replace 3 small spheres with larger ones. Ask yourself the same questions. Sketch your model(s).

2. Place 4 toothpicks (shortened), 1 at a time, into a 1½ inch sphere. The second is placed at an angle of 109.5° from the first, using the 109.5° angle template. The third is placed so that it is at 109.5° from each of the first two. You will have to rotate the sphere and do this by trial and error. The fourth is placed at 109.5° from each of the others. This is called a *tetrahedral* shape. (Is there more than one way to do this?)

Place a small sphere on each toothpick so that the spheres touch. This is a tetrahedrally shaped particle. Sketch your model.

In sequence, replace 1, 2, and 3 spheres on the tetrahedral shape, and at each point, determine how many uniquely different shapes can be made. Sketch each model.

3. Repeat the entire process above, but place 3 toothpicks at 120° from one another. This is a *trigonal planar* (some call it planar triangular) arrangement. (Is it possible to have the three toothpicks at 120° from one another and *NOT* have them in the same plane?) Again, replace 1 ball and 2 balls and ask the same questions as above. Sketch each model.

4. Using the trigonal planar shape from #3, place two additional toothpick segments into the central 1½ inch sphere at 90° to the plane of the first three (these are above and below the plane). Place a small sphere on each of the 5 toothpick segments. This is a *trigonal bipyramid* shape. In succession, replace one ball, then two balls, then three balls on this structure. Each time, determine the number of unique shapes you can produce, depending on which ball (or balls) you choose to replace. Sketch each unique model you have made.

Examine the trigonal bipyramid structure made with 4 small balls and one large one, all attached to the central sphere. Is there the same amount of space or room for the larger ball, regardless of where it is placed, or are some positions less crowded than others. If so, identify which positions are less crowded. Repeat this process using the structures with 2 large balls and 3 small ones.

5. Using a square planar shape as described in step #1, add two more toothpick segments, above and below the plane of the 4 balls and at 90° to the plane. Place 6 small balls around the central ball. This is an *octahedral* shape. In succession replace 1, then 2, then 3 small balls with larger ones, each time determining the number of unique shapes that can be made. Sketch all unique models.

As in step #4, assess the amount of crowding that the large spheres experience in each case, and determine the "best" way to arrange them to minimize this crowding with 1, 2, and 3 large replacement spheres.

DISCUSSION

The application of the above exercises to certain molecular shapes may not be obvious, but it very direct when one understands the basic principles. Electron pairs must have space to occupy, whether they are bond pairs or lone pairs. If there are 3 pairs of electrons around a central atom (remember that a double or triple bond counts as one pair), the electron pairs will adopt a trigonal planar shape, *regardless of how many pairs are bond pairs and how many are lone pairs.* When discussing the shape of the molecule, however, the lone pairs are invisible! This means that although methane, water, and ammonia all have 4 pairs of electrons around the central atom (see page 36), these molecules do not have the same shape. The four pairs are arranged in a tetrahedron in each case, but the invisible lone pair on NH_3 makes that molecule look like a compressed pyramid with the nitrogen atom at the top. The two lone pairs on the water molecule make the molecule look like a V (it is referred to as a "bent" molecule) with a bond angle very close to the tetrahedral angle of 109.5°. To visualize the differences in shape of these three, place 4 small spheres tetrahedrally around one larger one, as in step 2 of the procedure, and describe the shape. Then remove one small sphere (because the lone pair in NH_3 is still there occupying the position, but it is invisible). This is the shape of the NH_3 molecule. Remove a second small sphere (both lone pairs on water are still there but both are invisible). This is the shape of the water molecule.

Lone pairs occupy a little more room than bond pairs. In ammonia and water, this just means that the bond angles (H–N–H and H–O–H) in the molecules will be a little smaller than the tetrahedral angle of 109.5° to give the lone pairs more room. In NH_3, the H–N–H bond angle is about 107° and in water the H–O–H angle is about 105° because there two lone pairs crowding one another. This effect is most pronounced in the case of the trigonal bipyramid shape, where there is a significant difference in the available space in the equatorial position (3 in the plane) than in the axial position. The lone pairs will be found where they have the most room! With this in mind, predict and describe the shape of ClF_3, which has a trigonal bipyramid arrangement of electron pairs, where 3 are bond pairs and 2 are lone pairs.

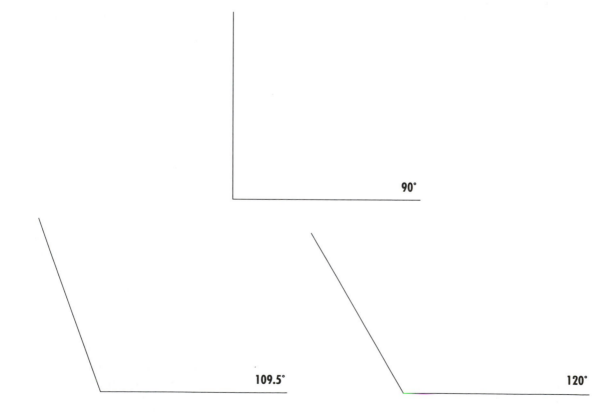

90°

109.5°   120°

PRE-LAB QUESTIONS

1. Why is a knowledge of the shape of a molecule helpful?

    _____

    _____

    _____

2. What chemist is responsible for developing the concept of electron dot diagrams for molecules?

    _____

3. Which "type" of electron pair requires more space around a central atom—a bond pair or a lone pair?

    _____

4. An atom in a molecule which is bonded to only one other atom is called a _____ atom, while an atom that is bonded to two or more atoms in a molecule is called a _____ atom.

5. Where in a molecule are lone pairs of electrons that *do not* affect the shape of the molecule?

    _____

Experiment 5
Data Sheet

Name _____

Date _____

MOLECULAR SHAPES ANSWER SHEET

SHAPE	NUMBER OF REPLACEMENT SPHERES	NUMBER OF UNIQUE SHAPES	SKETCH(S)
square planar	1	_____	
	2	_____	
	3	_____	
tetrahedral	1	_____	
	2	_____	
	3	_____	
trigonal planar	1	_____	
	2	_____	
trigonal bipyramid	1	_____	
	2	_____	
	3	_____	

Discuss the crowding and available room at each position in the structure with 1 & 2 replacement atoms.

_____

_____

_____

Octahedral   1   _____
                    2   _____
                    3   _____

Discuss the crowding and available room at each position in the structure(s) with 1, 2, and 3 replacement atoms.

_____

_____

_____

Based on the principles in the discussion section, predict how the electron pairs will be arranged, in each molecule or ion below, and describe the shapes of the molecule or ions. Use the spheres to make models to help you.

(a) TeF_4

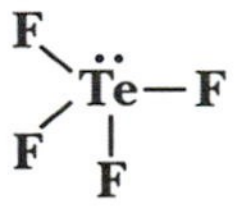

F
  \ ..
   Te—F
  / |
F   F

(b) SO_3

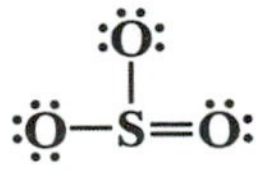

$:\ddot{O}:$
  |
$:\ddot{O}—S=\ddot{O}:$

(c) SO_3^{2-}

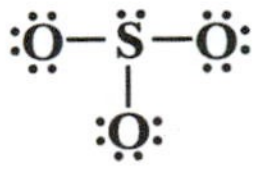

$:\ddot{O}—\ddot{S}—\ddot{O}:$
      |
    $:\ddot{O}:$

(d) ClO_4^-

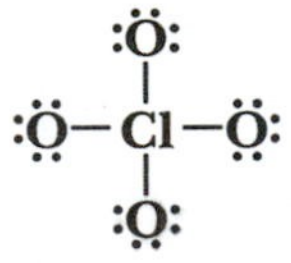

$:\ddot{O}:$
   |
$:\ddot{O}—Cl—\ddot{O}:$
   |
$:\ddot{O}:$

(e) ClO_3^-

$:\ddot{O}:$
   |
$:\ddot{O}—Cl:$
   |
$:\ddot{O}:$

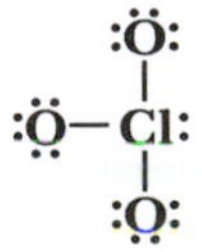

POST LAB QUESTIONS

1. For the shapes encountered in the lab, in which shape is there the greatest difference in the amount of room available for one large terminal atom, depending upon where it is positioned around the central atom?

   _____

2. In the shape identified in question 1, what position in this shape has the greatest amount of room?

   _____

3. In the two different shapes encountered having 4 terminal atoms around a central atom, what is the difference noted in structure when two atoms each of different sizes are placed around a central atom?

   _____

   _____

   _____

4. Can you suggest a reason for this difference?

   _____

   _____

   _____

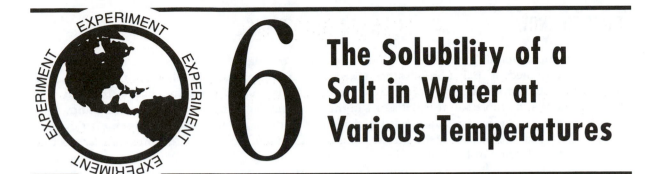

The Solubility of a Salt in Water at Various Temperatures

OBJECTIVE

To measure the solubility of a salt in water over a range of temperatures.
To represent salt solubility measurements graphically.

APPARATUS AND CHEMICALS

large test tube
split rubber stopper
110°C thermometer
10-mL graduated cylinder
ring stand

clamp
Bunsen burner
nichrome wire
distilled water
10 g $KClO_3$ or KNO_3

SAFETY CONSIDERATIONS

Protect your hands with a heavy towel when inserting a thermometer into the rubber stopper or cork.

Examine your test tube very carefully for small cracks before beginning the experiment. The heating and cooling cycles can put a great deal of stress on the glass, and small cracks invariably open up to large cracks during this process. In addition, the stirring must be done carefully so as to prevent breaking the bottom of the test tube. The lip of the test tube is also a particularly vulnerable area.

When using a burner or any open flame in the laboratory, be absolutely sure that loose hair and clothing do not come close to the flame. It is easy to forget this simple rule when concentrating on measurements.

The test tube can break under the heating and cooling cycles if it has an imperfection or imperceptible crack. This could cause a spattering of the hot solution. Wear your safety glasses at all times.

Be sure the gas is turned off completely when you are through with it since in a closed laboratory a small gas leak can build up to an explosive mixture.

Many salts, particularly chlorates and nitrates, can be unstable when very dry. Do not expose the dry salt to high temperatures, and under no circumstances attempt to grind up the dry salt (to dissolve it, for example) with an apparatus such as a mortar and pestle.

Mercury is a poison. If a thermometer is broken, have the instructor clean up the mercury as soon as possible. *Do not play with the mercury* because mercury is absorbed through the skin.

FACTS TO KNOW

When one material (the solute) dissolves in another material (the solvent) an interesting phenomenon is often noted. The first bit of solute added will dissolve, and additional amounts of solute added will continue to dissolve, up to a point. However, after a given amount of solute has been added, it will be observed that no more solute will dissolve in the solvent at that temperature, regardless of how much solute is added. The excess solute will simply collect at the bottom of the container. At this point, the solution is said to be saturated, and the quantity of solute dissolved *per 100 grams of solvent* is referred to as the solubility of that solute under the given conditions. The solubility of any substance in a particular solvent is a characteristic unique to that particular combination of materials and temperature. No two different solutes would have exactly the same solubility patterns in the same solvent.

In a saturated solution, the excess solute and the solution are said to be in dynamic equilibrium. The process of dissolution of the solute has not ceased, but rather solute particles are leaving the solution at the same rate that they are entering the solution (Figure 6-1). If the solution is saturated, and hence at equilibrium, its properties remain constant. If the solid is disappearing, the system is obviously not at equilibrium and the solution is not saturated. When a solution that is at or near saturation is cooled, and the solute becomes less soluble, the solution sometimes becomes supersaturated—that is, there is more dissolved solute than should normally be present. This occurs because the initial formation of crystals does not take place easily. Such solutions are said to be metastable, and when crystallization is induced, it will take place very rapidly to lower the concentration of the dissolved solute to the normal saturation level. Crystallization in these instances can often be induced by agitation of the solution or by adding a small "seed crystal" of the solute to the supersaturated solution. The vigorous stirring of the solution while it cools during this experiment is for the purpose of preventing supersaturation.

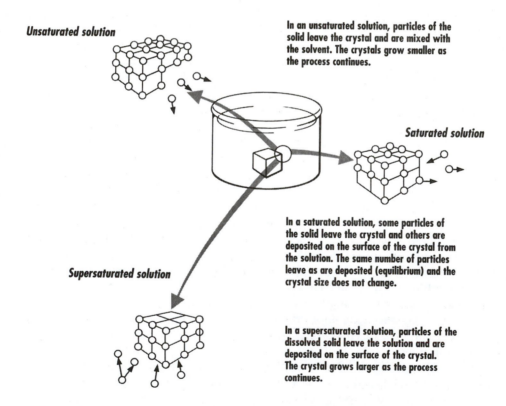

Unsaturated solution

In an unsaturated solution, particles of the solid leave the crystal and are mixed with the solvent. The crystals grow smaller as the process continues.

Saturated solution

In a saturated solution, some particles of the solid leave the crystal and others are deposited on the surface of the crystal from the solution. The same number of particles leave as are deposited (equilibrium) and the crystal size does not change.

Supersaturated solution

In a supersaturated solution, particles of the dissolved solid leave the solution and are deposited on the surface of the crystal. The crystal grows larger as the process continues.

Figure 6-1. Action of particles in unsaturated, saturated, and supersaturated solutions.

The solubility of a substance in any solvent depends largely on three classes of interactions. The first of these is the lattice energy of the solute (if the solute is a solid). Any factors which would make for a particularly high lattice energy would tend to work against solubility, since dissolving involves destroying (breaking down) the lattice.

For ionic substances, the two primary factors of importance are the charges of the ions and the sizes of the ions. Since the lattice energies depend on the product of the charges (positive charge × negative charge), as the charges go from 1 to 2 to 3 (either + or −), the lattice energy increases accordingly, by FACTORS of whole numbers (for example, a +2:−1 lattice (such as $CaCl_2$) has twice the electrostatic contribution to lattice energy as a +1:−1 lattice (such as NaCl) has, and a +3:−2 lattice (such as Al_2O_3) has 6 times the electrostatic contribution to lattice energy as a +1:−1 lattice. Compounds containing highly charged ions tend to be fairly insoluble as a result. Recall that in the general solubility rules, the "mostly soluble" anions almost all have −1 charges (halides, NO_3^-, ClO_4^-) and the "mostly insoluble" ions have higher charges (CO_3^{2-}, S^{2-}, PO_4^{3-}). The "mostly soluble" cations are all +1 (Na^+, K^+, NH_4^+). Sulfate (SO_4^{2-}) is the only common divalent "mostly soluble" anion, and it is quite large (see next factor discussed).

The second factor of importance for ionic compounds is the size of the ions. As the ions get larger, the electrostatic attraction for oppositely charged ions gets smaller, since electrostatic charges act as if they were at the center of the particle (the nucleus) and this gets farther from the opposing nucleus as the ions get larger. Again, from the solubility rules, one can see that the MOST HIGHLY SOLUBLE ionic solids contain univalent (−1) anions that are quite large (NO_3^-, ClO_4^-).

The two factors can be combined into one term that is known as "charge-to-size ratio" (charge/size). This can be thought of as a measure of the density of the charge on the particle. Note that a higher charge or a smaller size would make this term larger, and *vice versa*. Ionic compounds in which the charge-to-size ratio of the ions is high have high lattice energies, while those with large ions with low charges (such as nitrate) have very low lattice energies.

The second class of interactions of importance in solubility considerations is the interactions of the solvent molecules with one another. Here, the TYPE of interactions is important, as well as the strength of interactions. Solvents whose molecules are polar, such as water, have strong attractions for one another, due to the electrostatically charged "ends" of the molecules. Solvents whose molecules are non-polar (hydrocarbons such as hexane—C_6H_{14}—a component of gasoline, would be an example) interact well with one another but not with polar molecules such as water or with charged species such as the ions in an ionic lattice. This is the reason that table salt (NaCl) will dissolve well in water but not in gasoline, and the reason that gasoline and water will not mix well.

If only the lattice energies were of importance in determining how well ionic compounds dissolve in polar solvents, then this, as Voltaire's Candide would say, "would be the best of all possible situations." Solubility would be a simple thing to predict! Unfortunately, this is not the case. There is a third class of interactions that must be taken into consideration. This is the interaction of the solvent molecules with the solute particles. The problem "muddying the water" is that the same factors that make for high lattice energies also make for stronger solvent-solute interactions (which increase solubility). The higher the charge density (charge-to-size ratio) of the ions, the more strongly the molecules of water will associate with the ions (remember, this is an electrostatic interaction, also), and the greater the tendency for the ionic substance to dissolve. Thus, a higher lattice energy works against solubility, but the factors which work for a high lattice energy (charge density of the ions) also work toward greater solvent-solute interactions, which increases solubility. It is the balance of these interactions, plus the solvent-solvent interactions that determine the overall solubility patterns, which is why solubility must be experimentally determined and why only general trends can be observed.

One important feature of the solubility process is the energy change that accompanies dissolution of a solute in a solvent. The three classes of interactions discussed above combine to produce the overall energy change in the following way. Two of the three interactions produce endothermic (energy absorbing) energy flows. These are the lattice energy of the solid and the solvent-solvent interactions. Both of these are attractions that *must be overcome* for dissolving to take place, so *energy must be put in* to break these interactions. The lattice energy of the solute is always a *much larger* contributor to this total energy than are the solvent-solvent interactions, which are much weaker than lattice forces. The third type, solvent-solute interactions, are favorable interactions (or dissolution would not occur in the first place) and are akin to bond formation and hence *release energy*. The process of the solvent molecules forming attractions to solute particles is called *solvation* (hydration in the case where the solvent is water) and the associated energy change is called solvation energy (hydration energy when the solvent is water). It is the overall balance of the endothermic processes and the exothermic processes that determine whether a material will have an *exothermic* heat of solution or an *endothermic* heat of solution. Some substances release tremendous amounts of heat when dissolution takes place. For example, when NaOH dissolves in water, enough heat can be liberated to actually boil the water, if the proportions are correct. This is due to the very strong interactions of the hydroxide ion (OH^-) with the water molecules. On the other hand, when ammonium chloride (NH_4Cl) dissolves in water, the solution gets very cold. The hydration energy of the two ions is not nearly as large as the lattice energy, so the additional energy to break down the ionic lattice comes from the solvent (water) and reduces the temperature as a result. This process can be used to make the "cold packs" that an athlete uses when he/she sprains an ankle.

Because there is an energy change associated with the dissolution process (if this were not so, no process would take place!), and temperature is related to the energy of the materials present, the solubilities of all substances vary with temperature. The manner in which solubilities vary cannot be predicted, since there are competing factors that enter into the picture. Some materials get much more soluble as the temperature increases and some get only a little more soluble. Some get less soluble as the temperature increases. For some, the relationship between solubility and temperature is linear and for some it is a curve like an arm of a parabola. The pattern of solubility vs temperature for any solvent-solute combination cannot be predicted, but must be experimentally determined. It is most commonly observed, however, that solids become more soluble as the temperature of the solvent increases, and less soluble as the temperature falls. This observation will be used in this experiment to investigate the solubility changes with temperature for an unknown salt in water.

PROCEDURE

A graph must be prepared showing the solubility of the salt with temperature. *This may or may not be a linear (straight line) relationship.*

I. Assemble an apparatus consisting of a large (20 cm) test tube fitted with a notched cork, stirring wire, and $-10°C$ to $110°C$ thermometer. Included as a part of the apparatus should be a beaker large enough to keep the test tube standing upright.

II. Weigh the apparatus on the triple beam balance (1). It is good procedure to always use the same triple beam balance throughout an experiment if multiple weighings are to be done. All weighings in this experiment are to be done on the triple beam balance.

III. Add the appropriate amount of the unknown salt (from Table 6-1) to the apparatus and weigh the apparatus again. You will use only one of the salts.

TABLE 6-1

SALT	$KClO_3$	KNO_3
Weight of salt	4.5–5.0 mL	5.5–5.7 mL
1st quantity of water	10 mL	4 mL
2nd quantity of water	2 mL	2 mL
3rd quantity of water	8 mL	2 mL
4th quantity of water	5 mL	4 mL
5th quantity of water	5 mL	4 mL
6th quantity of water	10 mL	4 mL

IV. Using a graduated cylinder, measure out the initial quantity of deionized water (remember that for water, 1 gram ≈ 1 mL) and add it to the salt in the apparatus. Clamp the test tube as shown in Figure 6-2 and heat it with a Bunsen burner, stirring continuously as you heat. Use a "cool flame (it should NOT have the "double cone" associated with very hot flames). If the solution begins to boil before all of the salt has dissolved, add a small amount of water to the test tube with your squeeze bottle. Do not add more than is necessary to dissolve the salt. When all of the salt has dissolved, remove the flame and allow the solution to cool in the air. *YOU MUST STIR THE SOLUTION VIGOROUSLY ALL THE TIME IT IS COOLING!* Record the exact temperature when the solid begins to crystallize from the solution (when you first see crystals in the solution) (5a).

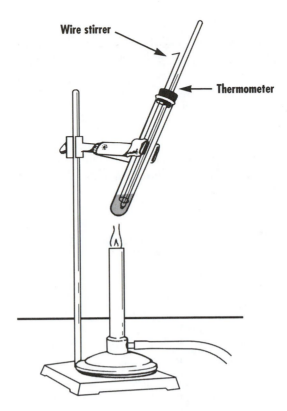

Figure 6-2. Set-up for a determination of solubility.

V. Cool the test tube to room temperature (you may use tap water), dry the test tube thoroughly, and weigh the apparatus (4). The weight of water present (5) is this weight, minus the weight of the apparatus + salt (2).

VI. Add a second measured portion of deionized water to the test tube and repeat the procedure described in numbers 4 & 5. The temperature of crystallization (7a) should be lower the second time.

VII. Continue this process of adding a measured quantity of deionized water, heating the solution to completely dissolve the salt, allowing the solution to cool (*WHILE STIRRING VIGOROUSLY*), determining the temperature of crystallization, and weighing the cooled apparatus, until a total of AT LEAST 6 weights and temperatures of crystallization have been recorded.

VIII. Discard the solution containing the dissolved salt in the container provided for that purpose.

CALCULATIONS

Prepare a table with columns headed "weight of water, weight of salt, solubility, and temperature of saturation."

In each case, the (weight of salt) is the (weight of apparatus + salt) − (the weight of apparatus). This value does not change throughout the experiment.

The weight of water present is always the *total amount in the test tube*, not just the amount added in any particular step. It is determined by subtracting the weight of the (apparatus + salt) from the weight of (apparatus + salt + water) for each determination. The *volume* of water *added* is not used in calculations.

The solubility of the salt at any temperature is calculated as (g salt)/(100 g water) as shown in the ratio:

$$\frac{\text{mass salt}}{\text{mass water}} = \frac{\text{solubility}}{100 \text{ g water}}$$

or

$$\text{solubility} = \frac{(\text{mass of salt}) * (100)}{(\text{mass of water})}$$

Prepare a graph of solubility (y-axis) vs temperature of crystallization (saturation) in °C (x-axis), and from the graph, predict the solubility the salt should have at temperatures of 70°C, 50°C, and 30°C. Read Lab Text 4 *Preparing and Using Graphs*—before preparing your graph. This graph is *not necessarily linear.*

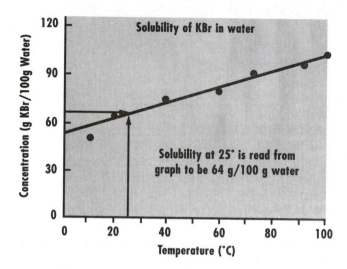

Figure 6-3. Sample plot of concentration of salt versus temperature.

PRE-LAB QUESTIONS

1. What three factors *associated with the solute and solvent* influence the solubility of an ionic substance in water?

   _____

   _____

   _____

2. How can crystallization be induced in a supersaturated solution?

   _____

   _____

   _____

3. What is the purpose of vigorously stirring the solution of dissolved salt as it cools?

   _____

   _____

   _____

4. What are the units on solubility—that is, how is the solubility of a substance reported?

   _____

   _____

   _____

Experiment 6
Data Sheet

Name _____

Date _____

SOLUBILITY OF A SALT IN WATER AT VARIOUS TEMPERATURES

(2) wt. of assembly + sample of salt   _____ g

(1) wt. of assembly   _____ g

(3) wt. of salt   _____ g

(4) wt. of assembly + salt + 1st addition of water   _____ g

                                                               Saturation

(2) wt. of assembly + salt   _____ g temperature

(5) wt. of water present   _____ g, (5a) _____ °C

(6) wt. of assembly + salt + water after 2nd addition   _____ g

(2) wt. of assembly + salt   _____ g

(7) wt. of water present   _____ g, (7a) _____ °C

(8) wt. of assembly + salt + water after 3rd addition   _____ g

(2) wt. of assembly + salt   _____ g

(9) wt. of water present   _____ g, (9a) _____ °C

(10) wt. of assembly + salt + water after 4th addition   _____ g

(2) wt. of assembly + salt   _____ g

(11) wt. of water present   _____ g, (11a) _____ °C

(12) wt. of assembly + salt + water after 5th addition   _____ g

(2) wt. of assembly + salt   _____ g

(13) wt. of water present   _____ g, (13a) _____ °C

(14) wt. of assembly + salt + water after 6th addition   _____ g

(2) wt. of assembly + salt   _____ g

(15) wt. of water present   _____ g, (15a) _____ °C

(16) salt used _____

DETERMINATION	WEIGHT OF WATER PRESENT	SOLUTION CONCENTRATION g SALT/100 g H_2O	SATURATION TEMPERATURE °C
1st	(5) _____	(5b) _____	(5a) _____
2nd	(7) _____	(7b) _____	(7a) _____
3rd	(9) _____	(9b) _____	(9a) _____
4th	(11) _____	(11b) _____	(11a) _____
5th	(13) _____	(13b) _____	(13a) _____
6th	(15) _____	(15b) _____	(15a) _____

Plot your data on the graph paper below. Draw a smooth curve to best fit the points. Plot temperature on the horizontal axis and g salt/100 g water on the vertical axis. Remember the importance of title, labels, and units of measure in making the graph (26).

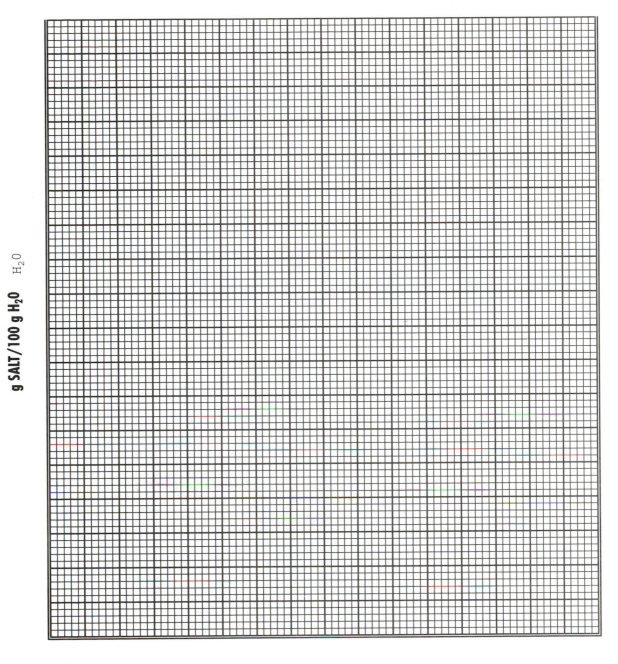

g SALT/100 g H$_2$O   H$_2$O

TEMPERATURE (°C)

Predicted solubilities from graph:

70°C _____

50°C _____

30°C _____

Solubility of the Salt _____ as a Function of Temperature

POST-LAB QUESTIONS

1. The total weight of water present at any time is not exactly equal to the volume added with the graduated cylinder, even though water has a density of 1.00 g/mL. Account for this difference.

    _____

    _____

    _____

2. Why is the weight of water present at any time obtained by weighing *after* the temperature of crystallization has been obtained, rather than immediately after the water is added to the test tube?

    _____

    _____

    _____

3. How would the graph change if a student inadvertently added one mL too much water in one step, then compensated by adding one mL less in the next step?

    _____

    _____

    _____

4. Using your graph, determine the solubility of your salt in water at 65°C (a) and at 45°C (b).

    (a) _____   (b) _____

7 Gas Law Experiments; The Law of Charles

OBJECTIVE

To examine how a change in temperature of a gas affects the volume of a gas and to use this effect to estimate the value of absolute zero.

To examine the effect of molecular weight on the velocity of a gas.

APPARATUS AND CHEMICALS

250-mL beaker
graduated cylinder
thermometer (220°C to 100°C)
wire gauze
Bunsen burner
15-inch glass tube (2 mm I.D.)
  sealed at one end
rubber band
ruler graduated in millimeters

glass tubing, 24 inches long
crushed ice
rock salt
mercury
ring stand
clamp
Q-tips
ammonium hydroxide (concentrated)
hydrochloric acid (concentrated)

SAFETY CONSIDERATIONS

When using a burner or open flame in the laboratory, be absolutely sure that loose hair and clothing do not come close to the flame.

Exercise caution when using glass tubing to avoid the possibility of injury to the hands through broken tubing.

Be sure the gas is turned off when you are through using the burner. In a closed laboratory a small leak can create an explosive mixture.

Exercise caution in the use of mercury. Mercury is very toxic and is absorbed through the skin; do not allow it to come in contact with your hands. In addition, mercury vapors are toxic.

FACTS TO KNOW

The temperature of a gas is a measure of the average kinetic energy of the molecules (or atoms) of a gas. Since kinetic energy is equal to $\frac{1}{2}mv^2$, where m is the mass of the molecule and v is the velocity, the temperature is related to how fast the molecules are moving. If the temperature rises (and the pressure remains constant), the molecules move more rapidly, and thus the gas occupies more space. If the temperature falls, the molecules move more slowly and the volume that they occupy is decreased. Charles' law states that at a constant pressure, the temperature (T) is linearly related to the volume (V) or

$$V = CT \qquad (1)$$

where the constant, C, is determined from the pressure (P) and the number of moles of gas (n):

$$C = \frac{nR}{P} \qquad (2)$$

in which R is a universal constant and has the same value for all gases,

$$0.08205 \; \frac{\text{liter atm}}{\text{mole K}}$$

Since the temperature is a function of the velocity of the molecules, when the molecules stop moving entirely, the temperature must be equal to zero. This zero is not the zero of any temperature scale in everyday use, but the temperature at which all molecules in the universe would cease to move. The volume-temperature relationship of equation 1 is the equation of a straight line; so if the volume of the gas is recorded at several different temperatures, the points can be plotted to give a straight line. This line, when extrapolated to $V = 0$, will give the value of this absolute zero of temperature.

PROCEDURE

Read the barometer and record the atmospheric pressure in mm of mercury pressure (1). If a barometer is not available, the instructor will post this figure. Obtain from your instructor a piece of thick-walled capillary tubing approximately 20–25 cm in length. The tube must be sealed at one end. This is done by holding one end of the tube in the hottest flame available from the lab burner (at the *outside edge* of the inner cone, near the bottom of the flame). Slowly rotate the tube with your fingers to keep the end from sagging as the glass softens and flows together. Your instructor will demonstrate this process for you. When the glass has melted and flowed together to close the end, have your instructor check the tube to see that it has been sealed properly. (Your instructor may provide pre-sealed tubes so that you may skip this portion of the procedure.)

Begin heating some water gently in a 250-mL beaker over your Bunsen burner. While the water is heating, take the glass tube that will be provided and pass it slowly through the burner flame (you should use a "soft" flame), *heating the sealed end first.* This will warm the air in the tube and remove any water vapor that may be present. (The glass will retain heat for some time, so be sure to hold the warm tube with a towel!)

While the glass tube is still warm, take it to the hood and immerse the open end in the beaker of mercury. As the air in the tube cools, the air contracts and the mercury is drawn into the tube. When a small "plug" of mercury (5 mm–10 mm) has been drawn into the tube, *gently* remove the tube from the mercury reservoir and allow the tube to cool to room temperature. The mercury plug should be about one-third down the tube.

Now fasten the glass tube to your thermometer with a rubber band so that the bulb of the thermometer is about at the midpoint of the trapped air column (Fig. 7-1). *Use caution in handling the glass tube, because if it is jarred the mercury will move or separate, and the volume of trapped air will change.*

Now fill the graduated cylinder with the water that has been heating. The side of the beaker should feel warm, and the water should be between 50°C and 60°C. Insert the tube and thermometer into the water and stir *very gently.* The water should completely cover the trapped air column. When the thermometer is stable and the mercury is no longer moving up, record the temperature (2), then withdraw the tube and quickly mark the exact position of the *bottom* of the mercury plug. Measure the height of the column from this point to the sealed end of the tube (3). (Be sure to measure to the end of the air passage, not to the end of the tube—the glass seal at the end has a volume of its own.)

Add a little tapwater to the warm water in the cylinder to lower its temperature 10 to 15 degrees (to between 30 and 40 degrees) and again insert the thermometer/glass tube assembly into the water. Allow the air column to come to equilibrium in this environment and record the temperature and height again (4 & 5).

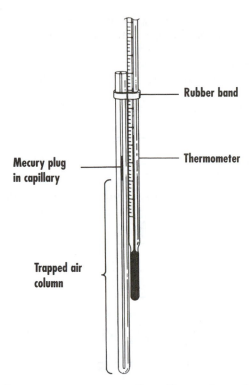

Figure 7-1. Correct arrangement of capillary tube and thermometer.

Pour out the warm water and fill the graduated cylinder with tap water (room temperature), and repeat the temperature and height measurements (6 and 7).

The fourth set of data points is obtained with a temperature between room temperature and zero degrees. This environment is attained by adding ice to about 100 mL of tap water in a beaker until the temperature of the water (after all of the added ice has melted) comes to about 10°C. This cold water is then poured into the graduated cylinder and the thermometer/tube assembly inserted and allowed to come to equilibrium as before. The temperature and the height of the air column are recorded as before (8 and 9).

The next set of data points are taken at the freezing point of water (0°C). A mixture of water and ice will come to zero degrees if sufficient ice is present in the water/ice mixture. This technique is used for calibrating thermometers at the zero degree mark. Pack the graduated cylinder loosely with ice (all the way to the top) and pour tapwater over the ice to about the 80 mL mark. Stir the mixture slowly with a stirring rod for about 3–4 minutes. Then insert the thermometer/tube assembly carefully into the mixture and slowly stir the mixture with the assembly until the air column has come to equilibrium (the mercury plug has stopped moving). Record the temperature and height as before (10 and 11).

The last points to be measured are at the temperature of an ice-alcohol mixture. Pack the cylinder with ice as before, but this time pour a 70% alcohol solution (prepared by your instructor) over the ice. Because solutions have a lower freezing point than pure water (this is the reason that salt melts ice on sidewalks in the winter), this mixture will come to a temperature between −5 and −10 degrees.

Insert the glass tube and thermometer in the cylinder and record the temperature and height (be sure to allow the temperature to become constant) as before (12 and 13).

These sets of values will be used to determine the value of absolute zero.
Read Lab Text 4—*Preparing and Using Graphs*—before preparing your graph. This graph should be linear, and the x-axis (temperature) should run from about −325°C to about +75°C to allow for extrapolation to h=0.]

Plot the data on a graph with height of the air column on the *y*-axis and temperature (°C) on the *x*-axis (10). Extrapolate the best line through the four points so that the line crosses the *x*-axis. Record the temperature where the line crosses the *x*-axis (14). This value represents absolute zero. Calculate your percent error using the known value for absolute zero (15). The capillary tube containing the mercury should be placed in the container provided in the hood.

PRE-LAB QUESTIONS

1. What are the dangers associated with the use of mercury in this lab?

   _____

   _____

   _____

2. What precautions must be taken when using mercury, as a result of the hazards identified in question 1?

   _____

   _____

   _____

3. How are temperatures below the ice point (0°C) produced in this experiment?

   _____

   _____

   _____

4. How is the universal gas law constant "R" related to the air trapped in the capillary tube in this experiment. Would it be different if pure nitrogen gas were used instead?

   _____

   _____

   _____

Experiment 7   **Name** _____
Data Sheet   **Date** _____

CHARLES' LAW

(1) Atmospheric pressure = _____

	Temperature (°C)	*Height of air column (mm)*
Warm water	(2) _____	(3) _____
Warm water	(4) _____	(5) _____
Tap water	(6) _____	(7) _____
Cold water	(8) _____	(9) _____
Ice-water mixture	(10) _____	(11) _____
Ice-salt mixture	(12) _____	(13) _____

(10) Plot of the length of a uniform air column vs. Celsius temperature

TEMPERATURE (°C)

(14) Value of absolute zero in °C (from graph)   _____

(15) Percent error (absolute zero)   _____

POST LAB QUESTIONS

1. What is the most likely source of error in determining the length of the air column at any given temperature in this experiment?

    _____

    _____

    _____

2. The value for absolute zero, expressed in Celsius temperature, is obtained by extrapolating the line obtained at temperatures close to room temperature into regions far removed from room temperature. What do you believe would be observed if you could actually reproduce these very low temperatures in the lab? That is, what would actually happen to the air column at these temperatures, and why?

    _____

    _____

    _____

8 Molecular Weight of a Volatile Liquid by Vapor Density

OBJECTIVE

To understand the direct relationship between the molecular weight of a gas and its vapor density and to use this relationship to determine the molecular weight of an unknown organic liquid.

APPARATUS AND CHEMICALS

balance (weighs to 0.0001 g)
125-mL conical flask
watch glass covers for 800-mL beakers
2.5 × 2.5″ squares of aluminum foil
3 × 3″ squares of aluminum foil
boiling chips
paper towels
ring stand
iodine crystals

ring
thermometer
wire gauze
burner with connecting hose
matches
800-mL beaker
50-mL (or 100-mL) graduated cylinder
15-mL samples of unknown liquid

SAFETY CONSIDERATIONS

  Some of the unknowns used in this experiment are flammable. Avoid having the unknowns close to an open flame.

  Hands should be protected when removing the flask from the hot water bath (see Fig. 8-3).

  Wear eye protection to shield against broken, splattered glass.

FACTS TO KNOW

A rather unique feature of the gaseous state is that ALL GASES BEHAVE IN THE SAME FASHION as far as their volumes, temperatures, and pressures are concerned. (This is not strictly correct, but the deviations are very slight under fairly moderate temperatures and pressures [above 0°C and less than 2 atmospheres pressure].) One consequence of the above statement is that one mole of ANY gas will occupy a particular volume under a particular set of conditions, and that one mole of ANY OTHER GAS will occupy THE SAME VOLUME under exactly those same conditions. This is somewhat surprising on the surface, since we commonly deal with masses, rather than moles, and one mole of one gas does not weigh the same as one mole of another gas. For example, one mole of O_2 (32 grams) under normal room conditions of 25°C and 1 atmosphere pressure will occupy about 24.5 liters of volume (fill a balloon to about 6.5 gallons). One mole of N_2 (28 grams), one mole of CH_4 (methane, 16 grams), one mole of SO_3 (sulfur trioxide, 80 grams) or 1 mole of UF_6 (uranium hexafluoride, 352 grams) WOULD ALL EXACTLY FILL THE SAME 24.5 LITER BALLOON! The obvious conclusion from these observation is that *the same number of gas molecules will always occupy the same volume under a given set of conditions.* [This is known as Avogadro's law.] This is true regardless of the type of gas present, or even if a mixture of gases is present. The same total number of molecules will be present in a mixture, but they will not all be the same kind. Since the volume of a gas varies greatly with temperature and pressure (but in a very predictable fashion—see Boyle's law and Charles' law topics in your textbook), a reference set of conditions, called Standard Temperature and Pressure (abbreviated {cleverly} STP) have been arbitrarily chosen. These are 1.00 atmospheres (760 torr) pressure and 0°C (273 K) temperature. Under these conditions, one mole of any gas occupies 22.4 liters. For this reason, 22.4 L is often referred to as the "molar volume" of a gas, but it is very important to remember that this number is only valid at STP (and is only valid for gases behaving "ideally," which refers to the small deviations mentioned above).

Since a known number of molecules of ANY gas will always occupy a certain volume under known conditions of temperature and pressure, and the mass of these molecules can be determined by simply weighing the container empty and again filled with the gas, the exact weight of a molecule (or, more accurately, a mole of the gas) can be obtained by dividing the mass of molecules by the number of moles of molecules. This gives a very powerful yet very simple way of determining the molecular weight of an unknown substance, IF THAT SUBSTANCE IS A GAS OR CAN BE CHANGED TO A GAS. Solids and liquids DO NOT CONFORM TO THIS BEHAVIOR PATTERN!

Because there is a direct relationship between the number of moles of gas molecules and the volume, temperature, and pressure of the container, and because this relationship is THE SAME FOR ALL GASES OR FOR MIXTURES OF GASES, a simple but powerful mathematical equation can be developed to show the relationship between these parameters. It is done in the following manner.

If the temperature of a gas is held constant, the volume of the gas is *inversely proportional* to its pressure (as pressure increases, volume decreases—we know this intuitively). Mathematically, this is $P*V = C$ (C is a constant). This observation is called Boyle's law.

If the pressure of a gas is held constant, the volume of the gas is directly proportional to the temperature of the gas. (As temperature increases, volume increases—we also know this intuitively, but the math equality holds ONLY IF THE TEMPERATURE IS EXPRESSED IN KELVIN DEGREES—ON THE ABSOLUTE SCALE—Celsius and Fahrenheit do not work.) Mathematically, this is $V/T = C'$ (C′ is a constant). This observation is called Charles' law (sometimes called Gay-Lussac's law).

If the temperature and pressure of a gas are held constant, then the only thing that can cause the volume of the gas to change is to change the number of molecules (practically speaking, the number of moles of gas). The volume would be directly proportional to the number of moles of gas present. Mathematically, this is $V/n = C''$ (C″ is a constant, and n = number of moles of gas). This observation is called Avogadro's law.

The three equations (Boyle's law, Charles' law and Avogadro's law) can be combined into one general equation, called the ideal (or perfect) gas law, and which describes the behavior of gases when any or all of volume, temperature, pressure, or amount of gas is allowed to change. This equation is:

$$(P*V)/(n*T) = R \quad \text{(R is a constant which combines C, C′, and C″).}$$

The most common way to see this equation is as $PV = nRT$. Since all gases conform to this equation, if the constant R is determined once, IT WILL BE THE SAME FOR ALL GASES! It can be easily calculated by remembering that at 273 K (0°C) and 1.00 atm pressure, 1.00 moles of gas occupy 22.4 liters.

$$(P*V)/(n*T) = R \quad \text{so} \quad [(1.00 \text{ atm})(22.4 \text{ L})]/[(1.00 \text{ mole})(273 \text{ K})] = R$$

This produces a value of $R = 0.0821$ L·atm/mol·K

The units on this value of R are critical! NOTICE THAT TO USE THIS VALUE OF R, ONE MUST USE VOLUME IN LITERS, TEMPERATURE IN KELVIN, AND PRESSURE IN ATMOSPHERES! One could determine values of R with a variety of units (but Kelvin temperature must ALWAYS be used).

For example, if pressure is used in torr and volume in mL, the correct value of $R = 6.24 \times 10^4$ mL·torr/mol·K. (Verify this for yourself!) This is not a different quantity, just the same quantity expressed in different units, like 1 ft and 12 in.

EXAMPLE CALCULATION

An example problem will be instructive in the use of the constant R. What is the molecular weight of a gas if a weight (w) of 5.0 g occupies a volume (V) of 8.5 liters when the temperature is 100°C ($T = 373$ K) and the pressure (P) is 1 atmosphere?

Rearranging we get

$$PV = nRT$$

$$M = \frac{wRT}{PV}$$

$$n = \frac{\text{mass}}{\text{mol. wt}} = \frac{w}{M}$$

$$PV = \frac{w}{m} RT$$

Substituting our values:

$$M = \frac{(5.0 \text{ g}) \, (0.082 \, \dfrac{\text{L atm}}{\text{mole deg}}) \, (373 \text{ deg})}{(1 \text{ atm}) \, (8.5 \text{ L})}$$

So

$$M = \frac{18 \text{ g}}{\text{mole}}$$

PROCEDURE

I. Using the analytical balance, weigh (together) a 125 mL conical flask and a small piece of metal foil (about 1½ inch square) (1). The flask should be very clean and dry on the outside. Do not clean the inside of the flask, since this will get the flask wet, and it must be dry on the inside!

II. Place approximately 15 mL of the unknown liquid into the flask and seal it as well as possible with the small piece of metal foil, Figure 8.1. If different unknowns are used, record the code letter for your unknown (2). Then, using a larger square of metal foil (3″ × 3″), cover the smaller piece as well as possible.

Aluminum foil (first cover)

Volatile compound

Figure 8-1. Loaded flask with first aluminum cover.

III. Place the flask in a water bath and cover the beaker with a large watch glass as shown in Figure 8.2. Heat the water bath until the water boils, and allow the water to boil for 10 minutes (Figure 8.3.) Then, remove the watch glass (using tongs!) and remove the conical flask BEFORE REMOVING THE HEAT FROM THE WATER BATH (IT SHOULD NOT STOP BOILING UNTIL THE FLASK IS REMOVED). While the flask is cooling to room temperature, measure the temperature of the boiling water (3).

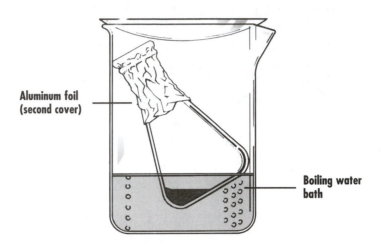

Figure 8-2. Loaded flask in beaker. Enough water is added to give a double-boiler effect. When the water is boiling, the entire flask will be in contact with either water at 100°C or steam at 100°C.

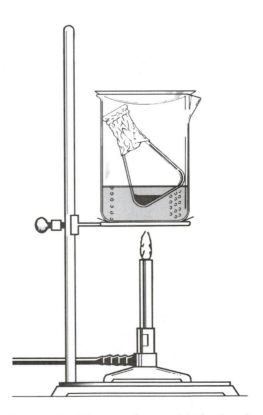

Figure 8-3. Ring stand assembly for heating water.

IV. Dry the flask thoroughly on the outside and allow it to cool to room temperature (at least 15–20 minutes). Remove the large piece of foil WITHOUT DISTURBING THE SMALLER PIECE. Dry the outside of the smaller piece as much as possible without removing it and weigh the flask and small foil on the analytical balance (4). There should be a very small amount of the unknown liquid in the flask at this point.

V. Remove the small piece of foil (SAVE IT) and rinse the flask with water (you may use tap water for this portion of the experiment since the density of tapwater is no different from that of deionized water) and then fill it completely (to the very rim) with tapwater. Weigh the filled flask (don't forget the small piece of metal foil) on the triple beam balance (it is much too heavy for the analytical balance) (5).

The barometric pressure in the room varies daily. (The weather report will give a number such as "30.2 and falling." What does this mean?) The pressure for the lab will be read by the instructor and the value will be written on the board. Record it (6).

CALCULATIONS

The basic equation to be used is $PV = nRT$, where n = number of moles. The number of moles of a substance present in a sample is given by n = (mass)/(Molecular Weight). Abbreviating molecular weight by M and substituting this for n in

$$PV = nRT \quad \text{gives} \quad PV = (mass)RT/(M)$$

This equation can be solved for M

$$M = (mass)RT/(PV)$$

Thus, the molecular weight of a gas (or substance that can be converted to a gas) can be determined if the volume, temperature, and pressure of a known mass of the substance can be measured. (mass) = the weight of the flask and foil after boiling minus the weight before boiling (3-1). The small amount of liquid remaining in the flask is the amount that it took to fill the flask as a vapor, under the existing conditions, while the flask was in boiling water. You started with a large excess of liquid—most escaped through the mouth of the flask as the flask sat in the boiling water bath and only the amount needed to fill the flask remained. When the flask cooled, this condensed to form a liquid, and the flask refilled with air. This is critical, since it was also filled with air the first time you weighed it!

R is the universal gas law constant mentioned above.

T is the temperature of the environment when the substance was a gas—that is, the boiling point of the water (which must be used in Kelvin).

P is the barometric pressure in the room, *in atmospheres*. The small foil, while it covers the mouth of the flask, cannot seal the flask, thus, since it is open to the atmosphere, its pressure must at all times equal the pressure in the room.

V is the volume of the flask. It is determined by subtracting the weight of the empty flask from the weight of the flask filled with water. This gives the weight of water necessary to completely fill the flask. Since the density of water is almost exactly 1.00 g/cm³, the volume of the flask is simply equal to the mass of water in the flask, converted to volume *and used in liters*.

Molecular weights of substances range from the smallest of 2.016 (for H_2) to values in excess of 500,000 (although these large ones cannot become gases under any normal conditions). The highest molecular weights for substances that are gases at or below the boiling point of water is about 352 (for UF_6), and most are well below this value. If your answer is outside this range, CHECK THE UNITS YOU HAVE USED!

PRE-LAB QUESTIONS

1. Give a concise statement of Avogadro's law

_____

_____

2. What are the numerical values for conditions at STP, and what is the value for R? INCLUDE UNITS!

_____

_____

3. What state(s) of matter DOES (DO) NOT obey the mathematical laws described in this experiment?

_____

_____

4. Why are two different pieces of metal foil used in this experiment, and what is the function of each?

_____

_____

5. How is the temperature of the gas in the flask maintained at a constant level and measured?

_____

_____

6. How is the pressure in the flask determined?

_____

_____

Experiment 8
Data Sheet

Name _____

Date _____

MOLECULAR WEIGHT OF A VOLATILE LIQUID BY VAPOR DENSITY

DATA AND ANSWERS TO QUESTIONS:

(1) Weight of 125-mL conical flask, aluminum square, and copper wire _____ g

(2) Code number for your unknown _____

(3) Temperature of boiling water _____ °C, _____ K

(3) Temperature of unknown sample in the flask _____ °C, _____ K

(4) Weight of cooled flask, contents and aluminum _____ g

(5) Weight of unknown sample that filled the flask at the

boiling point of water _____ g

(6) Atmospheric pressure _____ atm

(16) Volume of flask _____ mL, _____ L

(17) Molecular weight of unknown sample. _____ g/mole

Show your calculations below.

POST-LAB QUESTIONS

1. Three weighings are done in this experiment. The sequence of weighing is (1) empty flask + foil; (2) flask + foil *after* boiling water bath; (3) flask + foil + water to fill flask. Why is no weighing done when the unknown is initially added to the flask, *before* it is placed in the boiling water bath?

    _____

    _____

    _____

2. How would the calculated molecular weight of the unknown be affected if the flask were left in the boiling water bath for *too short a time* before being removed, and WHY? (too low, too high, or unchanged)

    _____

    _____

    _____

3. How would the calculated molecular weight of the unknown be affected if 20 mL of the unknown liquid had been placed in the flask at the start, rather than 15 mL, and WHY?

    _____

    _____

    _____

4. What must be true about the boiling point of the liquid in order for this method to provide an accurate molecular weight?

    _____

    _____

9 Determination of Molecular Weight by Depression of Freezing Points

OBJECTIVE

To determine the molecular weight of an unknown compound by measuring the extent to which a quantity of that compound depresses the freezing point of a solvent.

APPARATUS AND CHEMICALS

200-mm test tube
600-mL beaker
thermometer, 0.1°C (−10°C to +110°C)
25-mL graduated cylinder
ring stand

clamp
nichrome wire
rubber stopper
ice
cyclohexane

SAFETY CONSIDERATIONS

 Wear eye protection to prevent damage by the shattering of glass or the splashing of chemicals.

 Cyclohexane is flammable. Keep flames away.

 When inserting rubber stopper into a test tube, the tube may shatter and cut your hand. Protect your hand with a towel.

FACTS TO KNOW

Many years ago chemists noted that the dissolution of a nonvolatile compound (solute) in a liquid (solvent) lowered the vapor pressure of the solvent. Other properties related to the vapor pressure of the solvent (freezing point, boiling point, and osmotic pressure) were also changed. Such properties are known collectively as the *colligative properties* of a solution. At low solute concentrations, as long as the solute does not form ions, the change in the colligative properties is proportional only to the number of moles of solute particles present in a solution and not to the kind of particles dissolved.

There are many common examples of the use of these colligative properties. If a salt is added to a pot of boiling water, the water ceases to boil because the salt causes the water to boil at a higher temperature. Unless salt is added to an ice-water mixture when making ice cream in an ice cream freezer, the ice cream will not harden. The addition of salt lowers the temperature of the ice mixture sufficiently to freeze the ice cream. The addition of an antifreeze to the radiator of an automobile decreases the freezing temperature of the water in the radiator to a safe point. Salt sprinkled on icy streets and sidewalks melts the ice by depleting (through hydration) the water molecules in the vapor above the ice. The ice then sublimes to furnish more water molecules in the vapor state. Continuation of the process eventually melts the ice. The salt dissolved in the resulting solution prevents refreezing through depression of the freezing point. Application of salt to icy streets is effective only if the temperature is sufficient for ice to have an appreciable vapor pressure.

When working with colligative properties of solutions, it is convenient to express the solute concentration in terms of its *molality* (m), defined by the relationship

$$\text{Molality of A} = m_A = \text{Number of moles of A dissolved}/1000 \text{ g solvent}$$

or,

$$m = \frac{\text{moles of solute}}{\text{kg of solvent}}$$

The freezing point depression of a solution is given by $\Delta T_f = T°_f - T_f$ where $T°_f$ is the freezing point for pure solvent and T_f is the freezing point of the solution. In terms of concentration, the freezing point depression is

$$\Delta T_f = K_f m$$

where K_f is a constant characteristic of the solvent. Table 9-1 gives K_f values for some common solvents.

TABLE 9-1 K_f VALUES

Solvent	K_f
Water	1.86
Acetic acid	3.90
Camphor	20.0
Diphenyl ether	8.00
Cyclohexane	20.00

Colligative properties may be used to measure the molecular weight of an unknown substance. If a known amount of pure molecular solute is dissolved in a given amount of solvent, and the change in some property—freezing point, for example—is measured, and if an appropriate K_f value is known for the solvent, we can calculate the molality and hence the molecular weight. For the freezing point depression, the appropriate equations would be

$$\Delta T_f = K_f m = K_f \times \frac{\text{Moles of solute}}{\text{kg of solvent}}$$

Since

$$\text{Moles of solute} = \frac{\text{Grams of solute}}{\text{Molecular weight of solute}}$$

A substitution is made in the above equations to give

$$\Delta T_f = K_f \times \frac{\dfrac{\text{Grams of solute}}{\text{Molecular weight of solute}}}{\text{kg of solvent}}$$

Rearrangement gives

$$\text{Molecular weight of solute} = \frac{K_f \times \text{grams of solute}}{\Delta T_f \times \text{kg of solvent}}$$

You will be asked to find the molecular weight of an unknown solute using the last equation. The solvent to be used is cyclohexane, which has a melting point near 6.5°C and a K_f value of 20.00.

PROCEDURE

Numbers in parentheses refer to entry numbers on the data sheet.

I. FREEZING POINT OF CYCLOHEXANE Weigh an empty 200-mm test tube (1). Measure out about 25 mL of cyclohexane with a graduated cylinder and pour into the weighed test tube. Weigh the test tube and contents (2). Determine the weight of the cyclohexane from the two previous weights (3).

Insert a rubber stopper fitted with a stirrer and a thermometer (Fig. 9-1) into the test tube and clamp the assembly to a ring stand. Insert the test tube into a beaker containing a mixture of ice and water. Be sure the test tube is deep enough in the ice-water mixture to cover the contents of the test tube.

Stir the cyclohexane continuously and record the temperature every 30 seconds on the data sheet. Note the temperature at which solid first appears in the cyclohexane.

At the discretion of your instructor, the cyclohexane may be warmed in warm water, and the process repeated.

On the graph paper provided, plot temperature versus time and determine the freezing point of the cyclohexane, which is the point of change in slope of the line in the graph (4).

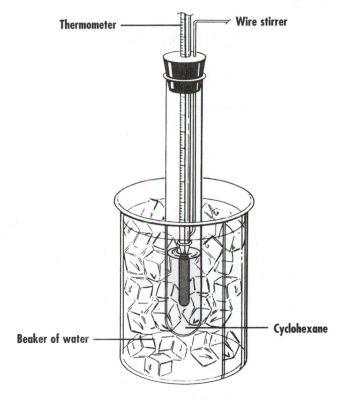

Figure 9-1. Apparatus for determination of freezing point depression.

II. MOLECULAR WEIGHT OF AN UNKNOWN SOLUTE Secure an unknown compound from the instructor. Weigh to the hundredths of a gram about 0.5 g of the unknown on a small sheet of weighing paper (5, 6, and 7). Carefully add the weighed solid to the test tube containing the cyclohexane.

Stir the mixture until all of the unknown has dissolved.

Lower the test tube into the ice-water mixture and begin stirring gently and continuously. Take temperature measurements every 30 seconds and record on the data sheet. If your solution does not freeze in the ice-water mixture, salt can be added to the ice-water mixture to lower its temperature.

Mark on your data where the temperature appears to become constant.

There may be some supercooling before freezing occurs. Supercooling can be avoided usually by steady stirring of the solution. Supercooling will be seen when a solution has to cool below its normal freezing temperature in order to form the first crystals. After the first crystals are formed, others form somewhat rapidly, and the heat given off causes the temperature to rise before continuing on its downward course.

Again, you may be instructed to repeat this procedure. If so, put the system in warm water until all solid melts. Then, return the test tube and solution to the ice-water mixture and record the temperature every 30 seconds.

On the graph paper, plot temperature versus time and determine the freezing point of the solution (the point of change in slope of the line) (8).

CALCULATIONS

Calculate ΔT_f, which is the difference between the freezing point of cyclohexane (4) and the freezing point of the solution (8). Put the answer for ΔT_f at (9).

Use the value for ΔT_f and 20.00 for K_f to calculate the molecular weight for your unknown (10).

EXAMPLE CALCULATION

$$\Delta T_f = 3.9°C$$
$$\text{Grams of cyclohexane} = 20.00 \ (0.02000 \text{ kg})$$
$$\text{Grams of unknown} = 0.49$$

$$\text{Molecular weight of unknown} = \frac{K_f \times \text{grams of solute}}{\Delta T_f \times \text{kg of solvent}}$$

$$= \frac{20.00 \times 0.49}{3.9 \times 0.02000}$$

$$= 126 \text{ (or 130 to proper number of significant digits with units of grams/mole)}$$

PRE-LAB QUESTIONS

1. List one important safety precaution to be taken when handling cyclohexane.

   _____

   _____

2. Describe two common examples where the freezing point lowering of water is encountered in everyday life.

   1. _____

   2. _____

3. Describe a precaution which should be taken when inserting a rubber stopper into a test tube.

   _____

4. What mathematical procedure is used to determine the freezing point of the solutions in this experiment?

   _____

   _____

Experiment 9   **Name** _____
Data Sheet   **Date** _____

DETERMINATION OF MOLECULAR WEIGHT BY DEPRESSION OF FREEZING POINTS

(2) Weight of test tube and cyclohexane   _____ g
(1) Weight of test tube   _____ g
(3) Weight of cyclohexane   _____ g
(4) Freezing point for cyclohexane from graph   _____ °C
(5) Weight of weighing paper + unknown   _____ g
(6) Weight of weighing paper   _____ g
(7) Weight of unknown   _____ g
(8) Freezing point for solution from graph   _____ °C
(9) ΔT_f (subtract 8 from 4)   _____ °C
(10) Molecular weight of unknown   _____

Temperature-Time Readings

Time (minutes)	*Temperature (°)* *cyclohexane*	*solution of unknown in cyclohexane*
0	_____	_____
$\frac{1}{2}$	_____	_____
1	_____	_____
$1\frac{1}{2}$	_____	_____
2	_____	_____
$2\frac{1}{2}$	_____	_____
3	_____	_____
$3\frac{1}{2}$	_____	_____
4	_____	_____
$4\frac{1}{2}$	_____	_____
5	_____	_____
$5\frac{1}{2}$	_____	_____
6	_____	_____
$6\frac{1}{2}$	_____	_____
7	_____	_____
$7\frac{1}{2}$	_____	_____
8	_____	_____
$8\frac{1}{2}$	_____	_____
9	_____	_____
10	_____	_____

Temperature as a Function of Time

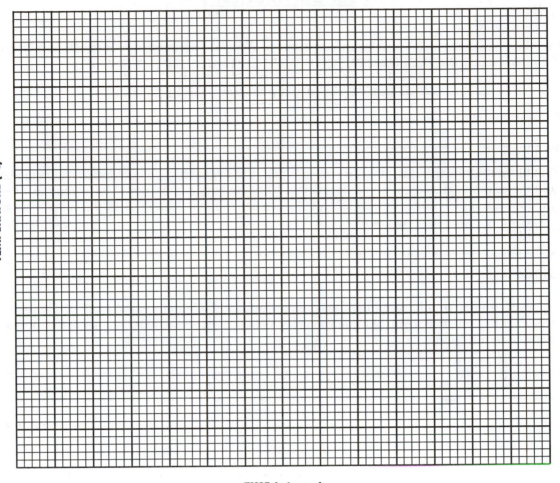

TEMPERATURE (°C)

TIME (minutes)

POST-LAB QUESTIONS

1. Cyclohexane solvent was chosen for this experiment. Would the accuracy of your results been as great if diphenyl ether had been chosen?

   _____

   Explain your answer. _____

   _____

2. In determining the freezing point of your solution, you might have observed super cooling about the time of the beginning of crystal formation. If so, the temperature would have gone to a lower initial value but then rather quickly increased to a more constant value. Should this initial low value be taken as the true melting point?

   _____

   Explain _____

3. Suppose a student in this experiment weighed the unknown solid only to the tenth of a gram accuracy rather than the hundredth of a gram accuracy. Could this cause a significant error in the molecular weight determination?

   _____

   _____

10 Separation of Dyes by Paper Chromatography

OBJECTIVE

To study the use of paper chromatography to separate mixtures of food color dyes.

APPARATUS AND CHEMICALS

Standards for Food, Drug, and Cosmetic
  Dyes (FD&C) Red #3, Red #40, Yellow
  #5, Blue #1
Food dyes (green, yellow, red, blue)
rubber bands
400-mL beaker, 600-mL beaker
sodium chloride
stapler
Saran wrap

Whatman No. 1 Chromatographic Filter Paper
  (9.5 × 16 cm)—3 pieces per student
Solutions of grape, lemon-lime, mountain
  berry Kool-Aid™
round toothpicks
scissors
pencil
ruler

SAFETY CONSIDERATIONS

Safety goggles should always be worn in a chemistry laboratory. No other special safety precautions are needed for this experiment becaus none of the materials used in this experiment are hazardous.

FACTS TO KNOW

The word *chromatography*, formed from the Greek words *chroma*, meaning color, and *graphein*, meaning to draw a graph or to write, was coined by the Russian botanist M.S. Tswett around 1906 to describe his process of separating mixtures of plant pigments. He separated the colored pigments in leaves by allowing a solvent mixture to carry the pigments down through a glass column packed with an insoluble material, such as alumina, silica, starch, or carbon. The term chromatography is now applied to any separation procedure employing the same principle as the method described by Tswett. There are four main types of chromatography: gas-liquid chromatography, column chromatography, thin layer chromatography, and paper chromatography.

All forms of chromatography employ the same general principle: A mixture of solutes in a *moving phase* passes over a selectively adsorbing medium, the *stationary phase*. Separation occurs because the solutes have different affinities for the stationary and moving phases. Solutes that have a greater affinity for the moving phase will spend more time in the moving phase and therefore will move along faster than solutes that spend more time in the stationary phase. There are a variety of ways of achieving a flow around or through a stationary phase.

In *column chromatography* small particles of the stationary phase are packed in a tube, and the moving phase (a liquid or a gas) flows around the particles confined in the tube; in *paper chromatography* the paper itself is the stationary-phase. In *gas chromatography* the mixture to be separated is vaporized and carried along a column containing a solid by using an inert gas such as helium (the carrier gas). The gaseous mixture is the mobile phase and the stationary phase is a finely divided solid coated with a high-boiling liquid (waxes, silicones) which is inside a column (usually stainless steel). The development of gas chromatography techniques and the sophistication of instrumentation make this method the most widely used technique for identification and separation of organic compounds.

In this experiment you will use paper chromatography to separate the components of food dyes. The cellulose structure of paper (Figure 10.1) contains a large number of hydroxyl groups that can form weak hydrogen bonds to water molecules, so the stationary phase can be regarded as a layer of water molecules hydrogen bonded to cellulose. If the solvent is water, the moving phase is also aqueous, but if a mixture of an organic solvent is used with water, the moving phase contains a high proportion of the organic solvent.

Figure 10-1. Cellulose structure of paper.

The solvents employed in paper chromatography must wet the paper so that the mobile phase will move through the paper fibers by capillary action. A solute in the mobile phase moves along with the solvent during the *development* of the chromatogram. As it moves, it undergoes many successive distributions between the mobile and stationary phases, the fraction of time it spends in the mobile phase determining how fast it moves along. If it spends all of its time in the moving phase, it will move along with the solvent front. If it spends nearly all of its time in the stationary phase, it will stay near the point of application.

After the chromatogram has been developed and the solutes on the paper have been located, the movement of the solution on the paper is mathematically expressed by the R_f value (called the *retention factor*), where R_f = *distance traveled by the solute divided by distance traveled by the solvent front*. The distances used to calculate R_f are shown in Figure 10.2. If all conditions of the experiment are kept constant, R_f values would be constant. Although variations in the paper and the temperature can cause slight changes in R_f, experiments run on the same day with the same materials should give constant R_f values. For this reason, your results for the standard food dyes should be reliable for identifying the dyes in food colors and Kool-Aid™ soft drink.

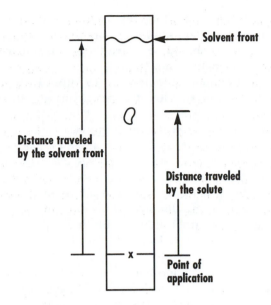

Figure 10-2. Distances used to calculate R_f factors.

PROCEDURE

Time can be saved in this experiment by getting part II ready while Part I is in the developing stages. Either 400 mL or 600 mL beakers can be used. The same developing solvent can be used for more than one chromatogram.

I. Preparations of Standard Chromatogram

A. Take one of the 9.5 × 16 cm pieces of Whatman No. 1 chromatographic filter paper and, with a pencil, draw a line parallel to the long dimension about 2 cm from the edge. Still using a pencil, put four small X's on the line, beginning 4 cm from the edge of the paper and spacing the marks about 3 cm apart.

B. Prepare 100 mL of the developing solvent by dissolving 0.1 g of NaCl in 100 mL of water. Add 25 mL of the 0.1% NaCl solution to a 400 or 600 mL beaker. Cover the beaker with a piece of Saran wrap and fasten with a rubber band.

C. Get a piece of small, scrap Whatman chromatographic filter paper and a toothpick and practice applying small (1- to 2-mm diameter) spots of one of the dyes on the paper. Put three spots over one another, allowing the spot to dry between each application. When you feel confident of this technique, take the 9.5 × 16 cm paper from (A) and lay it flat on a clean, dry paper towel. Use a clean toothpick for each of the standard food dyes and place *three overlying spots* of one of the dyes on each X. *Write in pencil under each spot the color and number of the standard.* Allow the spots to dry by gently waving the paper in the air. After they have dried, form the paper into a cylinder as shown in Figure 10.3 and staple the edges together, leaving a gap so that the edges do not quite meet.

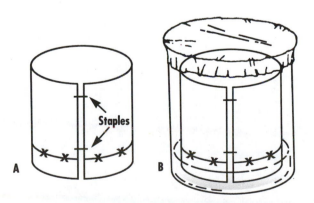

Figure 10-3. Form the chromatographic paper into a cylinder and staple as shown in A. Put the paper cylinder into the beaker containing the solvent, and cover with Saran wrap as shown in B.

D. Take the Saran wrap off the beaker from (B); put the paper cylinder in the beaker with the spots at the bottom, taking care that the paper does not touch the wall of the beaker. (**The spots should be above the surface of the developing solvent.**) Immediately replace the Saran wrap cover.

E. Allow the developing solvent to rise to within 1 cm of the top of the paper. This will take about 20 minutes. When this has occurred, remove the paper and place it on a paper towel.

F. *Mark the line reached by the developing solvent (it will be about 1 cm from the top of the paper) and the center of the spots as soon as possible after you remove the paper from the beaker. The marking must be done while the paper is still wet because the solvent and the spots keep moving.* Locate the densest part of the spot for each dye and draw a pencil line through it. Measure the distance in mm from the point of application to the densest part of the spot and record this distance on the data sheet (1)–(4). Also measure the distance from the point of application to the line reached by the developing solvent and record this distance (5). Calculate R_f for each dye (6)–(9).

II. Chromatogram of Four Food Dyes

Use a 400 mL or 600 mL beaker for your chromatogram and repeat Procedure I for the four food dyes. You will need to mark about *five overlying spots* to get a good chromatogram. Record the distances travelled by all spots (10)–(13) and the solvent (14) on the data sheet. Calculate R_f values for all the spots you identified (15)–(18). What standard food colors are present in each of the food dyes you tested? (19)–(22).

III. Chromatogram of a Commercial Food Product

Part III of the Procedure is performed in the same manner as Part II, except that concentrated solutions of Kool-Aid™ soft drink are used. Prepare a chromatogram using all three Kool-Aid™ flavors. You will need to use about *10 overlying spots* for each Kool-Aid color to get a good chromatogram. Record the distances travelled by all spots (23)–(25) and the solvent (26) on the data sheet. Calculate R_f values for all the spots you identified (27)–(29). What standard food colors are present in each of the Kool-Aid™ samples? (30)–(32)

PRE-LAB QUESTIONS

1. What is the mobile phase and the stationary phase in paper chromatography?

    _____

2. Explain how paper chromatography separates a mixture of dyes of different colors.

    _____

    _____

    _____

3. Calculate the R_f factor for the spot shown in Figure 2.

    _____

4. Suppose you are separating two different organic dyes with paper chromatography. Both dyes are water soluble. Dye #1 is more soluble than dye #2. Dye #1 also adheres less strongly to the filter paper being used for the stationary phase. A mixture of the dyes is dissolved in water. Spots are placed on a piece of filter paper, and the procedure described in Part I is followed. Which dye will have the larger R_f factor? Explain.

    _____

    _____

    _____

    _____

Experiment 10
Data Sheet

Name _____

Date _____

SEPARATION OF DYES BY PAPER CHROMATOGRAPHY

I.   STANDARD DYES

(1) Blue #1   Distance spot traveled   _____mm

(2) Yellow #5   Distance spot traveled   _____mm

(3) Red #3   Distance spot traveled   _____mm

(4) Red #40   Distance spot traveled   _____mm

(5) Developing solvent   Distance traveled   _____mm

Calculation of R_f values

(6) Blue #1

Blue #1 R_f value   _____

(7) Yellow #5

Yellow #5 R_f value   _____

(8) Red #3

Red #3 R_f value   _____

(9) Red #40

Red #40 R_f value   _____

II.   FOOD DYES

(10) Blue

Identify all spots by listing a color for each spot and the distance it traveled.

_____

(11) Green

Identify all spots by listing a color for each spot and the distance it traveled.

_____

(12) Red

Identify all spots by listing a color for each spot and the distance it traveled.

_____

(13) Yellow

Identify all spots by listing a color for each spot and the distance it traveled.

_____

(14) Developing Solvent: Distance traveled _____mm

Calculation of R_f values for all spots

(15) Blue

R_f value _____

(16) Green

R_f value _____

(17) Red

R_f value _____

(18) Yellow

R_f value _____

What food dye standards are present? List color and number for all the standards you found in the food dyes.

(19) Standards present in blue food dye: _____

(20) Standards present in green food dye: _____

(21) Standards present in red food dye: _____

(22) Standards present in yellow food dye: _____

III. KOOL-AID™ SAMPLES

(23) Grape

Identify all spots by listing a color for each spot and the distance it traveled.

_____

(24) Lemon-Lime

Identify all spots by listing a color for each spot and the distance it traveled.

_____

(25) Mountain Berry

Identify all spots by listing a color for each spot and the distance it traveled.

_____

(26) Developing Solvent—Distance traveled   _____mm

Calculation of R_f values for all spots

(27) Grape

R_f value _____

(28) Lemon-Lime

R_f value _____

(29) Mountain Berry

R_f value _____

What food dye standards are present? List color and number for all the standards you found in the Kool-Aid™ samples.

(30) Standard dyes present in Grape: _____

(31) Standard dyes present in Lemon-Lime: _____

(32) Standard dyes present in Mountain Berry: _____

POST-LAB QUESTIONS

1. Why is it important to mark the solvent front and the center of the dye spots at the same time?

    _____

    _____

2. What causes dye spots to be elongated rather than round?

    _____

    _____

3. Why are the Kool-Aid™ spots for dyes fainter than the spots observed for food dye mixtures in Part II?

    _____

    _____

84

Determination of the Empirical Formula of a Compound

OBJECTIVE

To learn how the weights of reactants and products in the oxidation of a metal can be used to determine the empirical formula of a metal oxide.

APPARATUS AND CHEMICALS

porcelain crucibles (2)
ring stands (2)
crucible holders (wire triangle) (2)
ring clamps (2)
Bunsen burners (2)

glass medicine dropper
tin foil
concentrated nitric acid
concentrated hydrochloric acid

SAFETY CONSIDERATIONS

 The Bunsen burner is used in this experiment. Care should be exercised in igniting and using the burner. Don't forget to turn the gas off after use.

 Splashed acid and the possibility of splattered, broken glassware warrant eye protection.

 Care should be taken to avoid touching the hot metal ring stand.

 The concentrated hydrochloric and nitric acids are very corrosive and will destroy skin and eyes. Eye protection must be worn and the acids must not be spilled or splashed on skin or in eyes. The reaction in this experiment gives off toxic oxides of nitrogen.

 These acids will destroy clothing on contact.

FACTS TO KNOW

The empirical formula (a formula derived by experiment) gives the relative *numbers of the different kinds of atoms* which are present in a compound. We can determine the relative weights of the different elements in a compound if we can synthesize the compound from its elements or analyze the compound to obtain the constituent elements. From a knowledge of the *atomic weights* of the different elements, we can determine the relative numbers of *atoms* (or moles of atoms) in a given weight of the compound. For example, a compound contains x grams of element X and y grams of element Y, then

$$\text{The number of moles of atoms of X} = \frac{x\text{ g}}{\text{g/mole X}}$$

$$\text{The number of moles of atoms of } Y = \frac{y\text{ g}}{\text{g/mole Y}}$$

The relative number of moles of atoms of the two elements would then be the ratio of X to Y in the chemical formula. In writing the empirical formula we convert the ratios to the relative numbers of atoms to a ratio of smallest whole numbers, and use the smallest whole numbers in writing the empirical formula.

A simple example will show how this is done in a specific case. A 1.000-g sample of iron is heated in air, during which time it unites with the oxygen of the air to produce 1.430 g of an oxide. What is the empirical formula of the oxide? We know that it must contain 1.000 g of iron for every 0.430 g of oxygen. Since the atomic weight of iron is 55.847 and that of oxygen is 15.9994, the relative number of atoms of these elements in this compound is

$$\text{moles of Fe atoms} = \frac{1.000\text{ g}}{55.847\text{ g/mole Fe}} = 0.0179\text{ mole Fe}$$

$$\text{moles of O atoms} = \frac{0.430\text{ g}}{16.00\text{ g/mole O}} = 0.0268\text{ mole O}$$

The ratio can be converted to a small whole number ratio by dividing the larger number by the smaller number:

$$\frac{\text{moles of O atoms}}{\text{moles of Fe atoms}} = \frac{.0268}{.0179} = 1.50$$

The ratio 1.50 to 1 is 3/2, so this is the mole (or atom) ratio of O/Fe in the compound. The empirical formula states this ratio simply as Fe_2O_3.

This experiment determines the weight of tin and oxygen in a given sample of tin oxide. From the weights obtained, an empirical formula is calculated. We shall oxidize *tin* with *nitric acid* and then heat the product to evaporate any unreacted nitric acid so that only a compound of tin and oxygen remains. The reaction can be abbreviated as

$$Sn \overset{HNO_3}{\rightarrow} SnO_x$$

where x is the key number if the empirical formula to be evaluated in this experiment. If we know the initial weight of tin used and the weight of the oxide produced, then the weight of oxygen present is given by the increase in the weight of the sample. From these data we can then proceed to determine the empirical formula of the tin oxide by dividing the weight of tin by its atomic weight (118.7 g/mole) and the weight of oxygen by its atomic weight (16.0 g/mole) as in the previous example.

PROCEDURE

Numbers in parentheses refer to entry numbers on the data sheet.

First you must clean and dry your two crucibles and mark them so that you can distinguish them. Your instructor will explain the marking technique used. After washing the crucibles, if they need it, and wiping them dry, place them in the crucible holder and heat them, first gently and then more strongly, with the Bunsen burner. Transfer them, using the crucible tongs, to a clean, dry heat-resistant surface to cool. When they are completely cool, weigh each crucible as accurately as you can and record the weights on your data sheet (1a and b).

Into each crucible place a piece of tin foil weighing about 1 g and then weigh the crucibles plus the foil as accurately as you can; record this weight on your data sheet (2a and b). Subtract to obtain the weight of tin used (3a and b).

The next part of the experiment must be done in a hood. Put both of the crucibles in the hood and add concentrated nitric acid dropwise to each. Observe the reaction cautiously after the addition of each portion of concentrated nitric acid and do not add the nitric acid so fast that it foams or splatters out of the crucible. After the tin has completely reacted with the nitric acid, the evolution of brown fumes will stop. When this stage has been reached, the crucible is to be placed in the crucible holder on the ring stand (in the hood!) and warmed very gently to evaporate the nitric acid.

When the evaporation is complete (as evidenced by a stage where no more brown fumes of nitric acid are given off) heat the crucible and its contents strongly. **Do not heat the crucible strongly before all of the nitric acid has been evaporated or the sample may splatter and throw your experimental results off.** After the crucibles have been heated strongly for about five minutes, allow them to cool to room temperature. Then weigh them as accurately as you can and record the weights on your data sheet (4a and b).

Subtract the crucible and tin weight (2a and b) from the weight of crucible plus tin oxide to get the weight of oxygen in the oxide (5a and b).

Divide the weight of tin by its atomic weight (118.7 g/mole) to get the relative number of moles of tin atoms (6a and b). Do the same for oxygen (16.0 g/mole) (7a and b). Reduce the ratio to a whole number ratio and record the empirical formula (8a and b).

The crucibles may be cleaned out by scraping out the loose tin oxide and, **in a working hood,** dissolving any that sticks to the crucible with a few drops of concentrated hydrochloric acid.

PRE-LAB QUESTIONS

1. Both concentrated nitric acid and concentrated hydrochloric acid are used in this experiment. They must be handled with extreme caution. Explain why and state at least 2 precautions to use.

    _____

    _____

2. Quantities of tin and tin oxide weighed in this experiment are relatively small compared to quantities you weight in other experiments. Why must these weighings be done accurately (at least to 3 significant figures)?

    _____

    _____

3. A Bunsen burner is used in this experiment. What precaution should be taken in using the burner if you have long hair?

    _____

    _____

Experiment 11
Data Sheet

Name _____

Date _____

DETERMINATION OF THE EMPIRICAL FORMULA OF A COMPOUND

	Crucible #1 (a)	Crucible #2 (b)
(2) Weight of crucible + tin	_____ g	_____ g
(1) Weight of crucible	_____ g	_____ g
(3) Weight of tin used (subtract 1 from 2)	_____ g	_____ g
(4) Weight of crucible + tin oxide	_____ g	_____ g
(2) Weight of crucible + tin	_____ g	_____ g
(5) Weight of oxygen in oxide (subtract 2 from 4)	_____ g	_____ g
(6) Relative no. of moles of tin atoms	_____	_____
(7) Relative no. of moles of oxygen atoms	_____	_____
(8) formula of the oxide formed	_____	_____

Show your calculations below:

POST-LAB QUESTIONS

1. Why must the crucibles be further heated to evaporate excess nitric acid after the tin has all reacted?

    _____

    What quantity on the data sheet would be incorrect if this were not done? _____

    _____

2. In the final heating of the crucible and contents, heating must be started slowly and continued carefully. What error in your results would occur if this were not done?

    _____

3. Why must the crucibles be allowed to cool to room temperature before weighing?

    _____

12 Determination of the Formula of a Hydrate

OBJECTIVE

To determine the mole ratio of water to barium chloride in the hydrate $BaCl_2 \cdot XH_2O$.

APPARATUS AND CHEMICALS

analytical balance (preferred)
crucible and cover
Bunsen burner
ring stand

wire or clay triangle
hydrated barium chloride
crucible tongs

SAFETY CONSIDERATIONS

 When working with an open flame be sure to protect hair and loose clothing. Also, be sure the open flame is not inadvertently placed under or near any part of the desk assembly. Double check to be sure the gas supply is completely turned off after you are through using the burner.

 Be sure that you do not burn yourself on the hot equipment. Hot porcelain, glass, or metal can burn you severely even thought it does not give the visual appearance of being hot.

 Barium chloride is a soluble salt of a heavy metal and, as such, is a poison. Clean up waste material and dispose of it as directed by your instructor. Wash your hands at the end of the laboratory period.

 Always wear safety goggles when doing laboratory work. This is especially important when heating chemicals close to your face.

FACTS TO KNOW

Many compounds combine with water to yield solid materials. In some of these the water is very firmly bound and cannot easily be driven off. For example, NaOH is formed from the reaction

$$Na_2O + H_2O \rightarrow 2\,NaOH$$

In other instances the water is rather weakly held in compounds called *hydrates* and can be driven off by heating the compound. Hydrated salts are good examples. Many of these hydrated salts are solids in which the salt is combined with a fixed number of water molecules. Examples include $CuSO_4 \cdot 5H_2O$, $CoCl_2 \cdot 6H_2O$, $NaBr \cdot 2H_2O$, $Na_2CO_3 \cdot 10H_2O$, $NiSO_4 \cdot 7H_2O$, $MgSO_4 \cdot 7H_2O$, and $CaSO_4 \cdot 2H_2O$. The temperature at which these hydrated salts readily lose water differs for each salt.

In this experiment we are going to determine the number of moles of water that are combined with one mole of barium chloride in the hydrate $BaCl_2 \cdot XH_2O$. When this material is heated, the following reaction occurs:

$$BaCl_2 \cdot XH_2O \overset{\Delta}{\to} BaCl_2 + XH_2O$$

If we can determine the weight of water lost when a known weight of $BaCl_2 \cdot XH_2O$ is heated, we can calculate the mole ratio of water to barium chloride from their gram formula weights. Using the gram formula weights and the equation

$$BaCl_2 \cdot XH_2O \overset{\Delta}{\to} BaCl_2 + XH_2O$$

relative weights $208.24 + X18 = 208.24 + (X \cdot 18)$

we know that

$$\frac{\text{Weight of } BaCl_2 \text{ produced (g)}}{208.24 \text{ (g/mole)}} = \text{Number of moles of } BaCl_2$$

and

$$\frac{\text{Weight of water lost (g)}}{18 \text{ (g/mole)}} = \text{Number of moles of } H_2O$$

To determine the number of moles of water (which is x) per mole of barium chloride, we simply carry out the division

$$\frac{\text{Number of moles of } H_2O}{\text{Number of moles of } BaCl_2} = \frac{x \text{ (moles of } H_2O)}{1 \text{ mole of } BaCl_2}$$

PROCEDURE

Numbers in parentheses refer to entry numbers on the data sheet.

The crucible and cover used in this experiment should always be handled with the crucible tongs to prevent moisture from your hands changing their weight. The crucible and cover should always be weighed at room temperature to avoid errors from the convection currents set up in air around a hot crucible.

Heat your clean dry crucible and cover with a Bunsen burner for one minute to drive off any traces of water. Let them cool, weigh them on your balance as accurately as you can, and record the weight (1). If you are using an analytical balance for weighing, place between 1 and 1.2 g of the hydrate sample in the crucible, replace the cover, and weigh the crucible plus contents (2). If you are using a triple beam balance, which does not weight as precisely as an analytical balance, you must use a sample size between 2 and 2.5 grams. The benefit of using an analytical balance is that it can weigh much smaller samples and this reduces the amount of material required for investigations. Put the crucible on a wire or clay triangle on a ring stand and open the cover so that water vapor can escape and you can observe the results of heating the material (Fig. 12-1).

Heat the crucible very gently with your Bunsen burner for about 5 minutes. Bring the flame closer to the crucible bottom and heat more strongly for another 5 minutes. Finally, put the Bunsen burner directly under the crucible and heat for 10 minutes with the tip of the blue cone of the flame.

Next, put the crucible cover completely over the crucible and allow the crucible and cover to cool for 10 minutes.

When the crucible and cover have cooled to room temperature, weigh them and record the weight (3). In order to insure that a constant weight has been obtained, put the crucible and cover back on the triangle and heat for another 5 minutes. Repeat the cooling and weighing steps and record the weight (4).

If this weight is lower that the first weight by more than 0.0005 g (0.05 g if a triple beam balance is used) then it is likely that all of the water was not driven off during the first heating, and more was driven off during the second heating. To insure that all of the water has, indeed, been driven off, the crucible must be heated, cooled, and weighed at third time. The weight should agree as indicated above. The heating, cooling, and weighing cycle should be repeated until two weighings are within the limits above. This is called "bringing the crucible to constant weight." If the weight *increases* on heating, it indicates that you are not cooling the crucible long enough, and either weight may be in error. This weight should be very close to the weight observed after the first heating treatment. Using the data you have obtained, calculate the number of moles of water per mole of barium chloride in your sample.

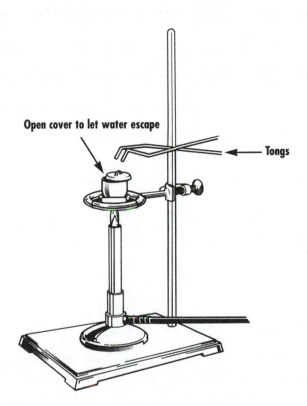

Figure 12-1. Heating the crucible.

PRE-LAB QUESTIONS

1. Barium chloride is poisonous. How should it be handled in the laboratory?

   _____

   _____

2. Why must barium chloride be disposed of in the designated container rather than by just dumping it into the sink?

   _____

   _____

3. What change does the starting material undergo in this lab when it is heated strongly?

   _____

   _____

4. How can you be certain that the desired reaction has gone to completion during the heating process?

   _____

   _____

Experiment 12
Data Sheet

Name _____

Date _____

DETERMINATION OF THE FORMULA OF A HYDRATE

EXPERIMENTAL DATA*

(1) Weight of empty, dry crucible and cover _____ g

(2) Weight of crucible, cover, and sample _____ g

(3) Weight of crucible, cover, and sample after initial-heating _____ g

(4) Weight of crucible, cover, and sample after second heating _____ g

CALCULATIONS

(2) Weight of crucible, cover, and sample _____ g*

(1) Weight of crucible and cover _____ g

(5) Weight of hydrate taken (subtract 1 from 2) _____ g

(4) Weight of crucible, cover, and sample after heating to constant weight _____ g

(1) Weight of crucible and cover _____ g

(6) Weight of anhydrous sample (subtract 1 from 4) _____ g

(7) Weight of water lost (subtract 6 from 5) _____ g

(8) Number of moles anhydrous $BaCl_2$ obtained

$$= \frac{\text{(Answer 6)}}{208.24} = \text{_____ moles}$$

(9) Number of moles of water lost

$$= \frac{\text{(Answer 7)}}{18} = \text{_____ moles}$$

(10) Mole ratio of $H_2O/BaCl_2$ (9 divided by 8) _____

(11) X in $BaCl_2 \cdot XH_2O$ _____

* Some numbered items are repeated to aid the student with the flow of thought in making the calculations.

POST-LAB QUESTIONS

1. Weighing must be done in this experiment after crucibles have cooled to room temperature. What error would occur if this were not done?

   _____

2. The procedure in this experiment instructs you to heat the crucible and contents again after the first heating, followed by cooling and reweighing. Why is this necessary?

   _____

3. When heating the barium chloride hydrate, you must heat gently at first, then more strongly. What type error could occur if you heated the barium chloride hydrate strongly at first.

   _____

4. What advantage does an analytical balance have over a triple beam balance for this lab?

   _____

   _____

13 Gravimetric Determination of Sulfate

OBJECTIVE

To illustrate an important analytical technique by determining the amount of sulfate present in a water-soluble mixture.

APPARATUS AND CHEMICALS

150-mL beaker
400-mL beakers (2)
gravity funnels (2)
stirring rods (2)
ring stand with ring (2)
wire gauze (2)
porcelain crucibles (2)
watch glass
sulfate unknown
desiccator

tongs
balance (± 0.001 g or better)
filter paper (Whatman #40 or #42-ashless)
Bunsen burner
0.3 M $BaCl_2$
concentrated HCl
$-10°C - 110°C$ thermometer
0.2 M $AgNO_3$
clay triangles (2)

SAFETY CONSIDERATIONS

 When using a burner in the laboratory, be absolutely sure that loose hair and clothing do not come close to the flame.

 Exercise caution to be sure that no hot materials splatter into the eyes while heating the samples. Wearing eye protection will insure your safety.

 Porcelain crucibles may be very hot without giving the appearance of being hot. Handle at all times with tongs to avoid burns.

 Be sure the gas is turned off when you are through with the burner since in a closed laboratory a small leak can create an explosive mixture.

 Hydrochloric acid can burn your skin, damage your eyes, and destroy your clothes. Wash off the hydrochloric acid with much water.

FACTS TO KNOW

Gravimetric analysis, an important technique of analytical chemistry, is concerned with *how much* of a component is in a given sample of material. A gravimetric analysis involves the following steps:

1. Accurately weigh samples of the substance to be analyzed.
2. Dissolve the samples in a suitable solvent—water, usually.
3. Add a chemical to the solution that will form a precipitate with the desired component.
4. Separate this precipitate from the liquid, normally by filtration.
5. Dry the precipitate.
6. Weigh the precipitate.
7. Calculate the percentage of the desired component from the weight of precipitate and of the sample.

Although this gravimetric method is applicable to many substances, the technique can be effectively demonstrated by analyzing for the sulfate (SO_4^{2-}) content of a water soluble salt mixture. The SO_4^{2-} is precipitated as solid barium sulfate ($BaSO_4$) by the addition of barium ions (Ba^{2+}). The barium sulfate precipitate is sparingly soluble in pure water, with only 2.4×10^{-3} g (.0024) dissolving in one liter of water at 25°C.

The solution is kept slightly acid to avoid the precipitation of other substances that might form in neutral or alkaline solutions. The equation for the precipitation is:

$$Ba^{2+} + SO_4^{2-} \xrightarrow{H^+} BaSO_4$$

The $BaSO_4$ can be collected by simple gravity filtration with filter paper, with subsequent heating of the residue until all traces of the paper have burned away, leaving only the $BaSO_4$ precipitate. This heating is done in a porcelain crucible that can withstand very high temperatures. The $BaSO_4$ is stable in the flame from a Bunsen burner.

This method of determining SO_4^{2-} is a very accurate method and can yield good results in the hands of a beginner.

This experiment also serves to demonstrate the concept of *stoichiometry.* The word literally means "the measurement of amounts of elements." As can be seen from the equation above, the Ba^{2+} and SO_4^{2-} react in a one-to-one ratio. However, an excess of Ba^{2+} is added to insure that essentially all of the SO_4^{2-} will precipitate. Since the quantity of SO_4^{2-} in $BaSO_4$ can be determined from the formula, the quantity of SO_4^{2-} in the precipitated $BaSO_4$ can be determined by using the simple ratio

$$\frac{\text{grams } SO_4^{2-}}{\text{grams } BaSO_4} = \frac{\text{Formula wt. } SO_4^{2-}}{\text{Formula wt.. } BaSO_4}$$

$$\text{grams } SO_4^{2-} \text{ in sample} = \text{grams } BaSO_4 \times \frac{\text{Formula wt. } SO_4^{2-}}{\text{Formula wt. } BaSO_4}$$

$$\text{grams } SO_4^{2-} \text{ in sample} = \text{grams } BaSO_4 \times \frac{96.00}{233.3} = \text{grams } BaSO_4 \times 0.4115$$

The precipitate of $BaSO_4$ as it is initially formed is very finely divided, so the solutions must be "digested." This means that the precipitate must be conditioned to increase the size of its particles so they will not pass through the filter paper. Digestion is accomplished by heat and/or time. The solutions can be digested effectively by heating them for one hour or by letting them sit for several days. The heat and/or time will cause (or allow) the particles to grow into larger particles. They then can be effectively collected on the filter paper.

Barium sulfate is a compound that has medical importance. While barium is a reasonably toxic metal ion, barium sulfate can be safely ingested (swallowed) because it is so very insoluble that it cannot dissolve in the aqueous stomach or intestinal fluids and simply passes through the intestines and is excreted, much like a rubber band would be. This makes it valuable for medical purposes, because barium, like lead, is opaque to X-rays. A common medical procedure for persons having gastro-intestinal problems such as diverticulitis, Crohn's disease, or colitis is to drink a barium sulfate "milkshake" and shortly thereafter have X-rays taken of their intestinal tracts. The insoluble barium sulfate coats the intestines and causes them to show up on the developed X-ray photographs. The physician can then identify swollen, blocked, or inflamed regions for treatment. The barium sulfate then works its way through the intestinal tract without dissolving and is excreted. Thus, the same property of insolubility that allows for the analysis of sulfate in an unknown makes barium sulfate a valuable medical tool.

PROCEDURE

This is a two-week experiment! *It is very important to read it thoroughly in advance so you will get the proper things done in the first week.*

The percent of sulfate in the compound will be determined by precipitating the sulfate as barium sulfate ($BaSO_4$) and weighing the precipitate.

Clean and rinse two 400-mL beakers. The experiment is to be done in duplicate. Accurately weigh out on a previously weighed watch glass (3 and 4) approximately 0.8 g to 0.9 g (do not exceed 0.9 g) of the unknown salt (1 and 2). The weight of the sample is obtained by difference (5 and 6). Transfer each sample to one of the beakers (mark beakers #1 and #2). Rinse the watch glass with a small amount of deionized water, collecting the rinsings in the beaker containing the sample. Add 200 mL of deionized water to each beaker. Add 6 mL of concentrated HCl solution. Stir until the sample is dissolved. A clear solution should be obtained. If the stirring rod is removed from the beaker, rinse the rod with deionized water, collecting the washings in the beaker.

Heat both solutions almost but not quite to boiling. Hold at this temperature by adjusting your burners.

Place a clean stirring rod in each beaker. Measure out 25 mL of 0.3 M $BaCl_2$ solution. Add a few drops of the $BaCl_2$ solution to one beaker, stirring the hot solution as the $BaCl_2$ is added. Then while stirring, *SLOWLY* add the remainder of the 25 mL of $BaCl_2$ solution to the beaker. Leave the rod in the beaker. Repeat the process with the second beaker.

The precipitate formed is very finely divided, so the solutions must be "digested." The solutions can be effectively digested by heating them for one hour or letting them sit for several days. Since filtration of the digested precipitate requires at least one hour, you should check the laboratory time to see which method to use. After the precipitate has been digested and has settled, check for completeness of precipitation by adding (without stirring) a drop of the $BaCl_2$ solution to each beaker. Note whether more $BaSO_4$ forms. If additional precipitate forms, add 1 mL of $BaCl_2$ solution while stirring. Allow precipitate to settle and check again for completeness of precipitation.

Mount your funnel support on a stand. Wash and rinse two funnels. Obtain two discs of #42 Whatman "ashless" filter paper from the laboratory instructor. (The filtration can be done with #40 filter paper if the sample has been "digested." This allows much faster filtration.) "Ashless" paper means that when the paper is burned practically noresidue remains. The discs you use produce on the average an ash of 0.0001 g (negligible for our purposes) per disc. A few minutes before the end of the digestion period, fold the papers in the same manner as your instructor demonstrates. Place the papers in the funnels. Moisten with deionized water and fit to the funnel with your clean fingers. A good fit will reduce the time it takes to filter the solution. Place receptacles under the funnels to receive the filtrate.

Transfer the precipitate to the filter paper. Your laboratory supervisor will demonstrate this procedure.

The filtration process is quite slow, since filter paper with very small pores must be used to catch the very finely divided precipitate particles. This procedure will be speeded up *significantly* if the solution is kept hot the entire time that the solution is being filtered. You must use care in handling the hot beakers, but the care will be rewarded with a shorter time for the filtering.

The precipitate must be washed free of any $BaCl_2$ it might contain. Heat a small beaker (125 mL) of *deionized* water. Add hot water slowly to the funnel, using your stirring rod to direct the water to the edge of the filter paper. After the first washings have run through the funnel, place a clean test tube under the funnel and wash a second time. To these washings, add a drop of silver nitrate ($AgNO_3$). If a white precipitate forms, the precipitate will have to be washed again and the washings tested. When all the $BaCl_2$ has been removed from the precipitate of $BaSO_4$, no precipitate of AgCl will form in the washings.

At the same time that you are precipitating the $BaSO_4$, you should prepare two porcelain crucibles for use at this point. This procedure is termed "bringing the crucibles to constant weight," and preparation consists of the following steps:

1. Wash thoroughly.
2. Place in wire triangles on ring stands.
3. Heat vigorously for ten minutes.
4. Allow to cool on triangle for one minute.
5. Transfer to desiccator by means of crucible tongs and allow to cool to room temperature, about one-half hour. (The desiccator must be charged with dried $CaCl_2$, which will be provided.)
6. Weigh and record weight (7 and 8).
7. Place on wire triangle and heat vigorously for five minutes.
8. Repeat steps 4, 5, and 6 (9 through 12, if needed).
9. Continue as in steps 7 and 8 until successive weighings agree within 0.001 g.

Place the folded filter paper plus the precipitate in the crucible. The instructor will show you how to remove and fold the filter paper. Place the crucible on a wire triangle supported by an iron ring. With a low colorless flame, gently heat the crucible by "wiping" the crucible with the flame until the filter paper and its contents are dry. Increase the temperature of the burner and wipe the crucible with the flame, continuing to heat in this manner until the paper has been charred completely. If the paper catches fire, stop heating until the flame goes out, then continue heating. Once the paper has been completely charred, the crucible should be heated with the most intense flame your burner can produce. *The distance from the top of the burner to the bottom of the crucible should be approximately 2 inches* (the innermost core of the flame will be about three-fourths of an inch high). Continue to heat until all of the carbon has been oxidized (all the black has disappeared). When the black has disappeared, stop the heating, allow the crucible to cool for one minute, transfer it to the desiccator, and allow it to cool to room temperature. Weigh and record weight (13 and 14). Heat again for five minutes with a very hot flame, cool as before, and weigh. Continue this process until two successive weights agree within 0.001 g.

Calculate the weight of the precipitate (21 and 22) and then calculate the percent of sulfate in the unknown salt (23 and 24). Report the average value (25). The gravimetric factor for SO_4^{2-} in $BaSO_4$ is 0.41153. The method of calculation is shown below:

$$\% \text{ sulfate} = \frac{\text{Wt. of } BaSO_4 \times 0.4115}{\text{Wt. of unknown sample}} \times 100\%$$

PRE-LAB QUESTIONS

1. This experiment is a gravimetric analysis. In a gravimetric analysis, what piece of general laboratory apparatus is used a number of times to collect important numerical data?

    _____

2. What becomes of the filter paper used to collect the barium sulfate precipitate?

    _____

3. What is meant by the term "digesting" a precipitate, and how is it done?

    _____

    _____

4. What is the purpose of adding the hydrochloric acid to the reaction mixture before the precipitation of the barium sulfate takes place?

    _____

    _____

5. What medicinal purpose does barium sulfate serve?

    _____

    _____

Experiment 13
Data Sheet

Name _____

Date _____

GRAVIMETRIC DETERMINATION OF SULFATE

	Sample #1	**Sample #2**
Weight of watch glass and sample	(1) _____ g	(2) _____ g
Weight of watch glass	(3) _____ g	(4) _____ g
Weight of sample	(5) _____ g	(6) _____ g
Weight of crucible	(7) _____ g	(8) _____ g
	(9) _____ g	(10) _____ g
	(11) _____ g	(12) _____ g
Weight of crucible and precipitate	(13) _____ g	(14) _____ g
	(15) _____ g	(16) _____ g
	(17) _____ g	(18) _____ g
	(19) _____ g	(20) _____ g
Weight of precipitate	(21) _____ g	(22) _____ g
Percent SO_4^{2-}	(23) _____ g	(24) _____ g
Average % SO_4^{2-} in compound	(25) _____	

POST-LAB QUESTIONS

1. What was the purpose of heating the porcelain crucibles a second or third time during the initial weighing process?

    _____

    _____

    _____

2. How can one be sure that all of the filter paper was burned away during the heating of the filter paper plus precipitate step?

    _____

    _____

3. What would have been the effect on the final value for the per cent sulfate in the unknown if the precipitate had not been washed with deionized water before the paper was burned away, and why would the value have changed in this manner?

    _____

    _____

    _____

4. What would have been the effect on the final value for the per cent sulfate in the unknown if the precipitate had not been adequately digested prior to filtration?

    _____

    _____

    _____

5. Would you have known that the precipitate had not been adequately digested? If so, how would you have known?

    _____

    _____

Calorimetry: The Determination of the Heat Effects that Accompany Chemicals Reactions

OBJECTIVE

To construct a simple calorimeter and learn how it is used to determine the heat changes which accompany chemical reactions in solution.

APPARATUS AND CHEMICALS

12-oz Styrofoam cups (2)
thermometer with 0.10°C or 0.20°C divisions
balance
cap for Styrofoam cup
400-mL beaker
250-mL beaker

100-mL graduated cylinder
Bunsen burner
ring stand and ring
wire gauze
3.0 M hydrochloric acid
3.0 M sodium hydroxide

SAFETY CONSIDERATIONS

Use care in handling glass apparatus. Glassware is particularly slippery when it is wet, and shattered glass can cause severe cuts.

When using a burner or any open flame in the laboratory, be absolutely sure that loose hair and clothing do not come close to the flame. It is easy to forget this simple rule when concentrating on measurements.

The acid and base solutions used in this lab can each cause permanent eye damage if splashed into the eyes. Wear your safety glasses at all times. If some reagent does get into your eyes, immediately call for help from your instructor and begin flushing your eyes with cold water.

Be sure your gas is turned off completely when you are through with it since in a closed laboratory a small gas leak can create an explosive mixture.

Neither the acid nor the base is concentrated enough to cause immediate burns, but they should be washed off *immediately* with cold running water, followed by soapy water, if any splashing or spilling occurs. This applies to fabrics as well as to skin.

FACTS TO KNOW

We can determine the heat effects which accompany a chemical reaction if we can carry out the reaction in a container which prevents heat exchange with the surroundings. When a container is insulated effectively, it will gain or lose heat to its surroundings very slowly. A container of this sort is called a calorimeter. A very simple calorimeter can be constructed from two Styrofoam cups, one placed inside the other. In addition to the cups we need a thermometer to measure the temperature change which accompanies the reaction and a stirrer to insure the complete mixing of the reacting solutions. A simple calorimeter of this sort is shown in Figure 15-1.

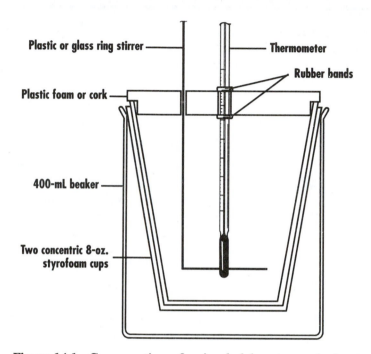

Figure 14-1. Cross section of a simple laboratory calorimeter.

The calculations of the heat involved require that we measure the temperature of the calorimeter and its contents *before* and *after* the reaction. The temperature will increase if the reaction gives off heat (an exothermic reaction). If heat is absorbed by the reaction (an endothermic reaction), the temperature will decrease. The amount of change in the heat content of the system (its enthalpy change, or ΔH) is equal to the product of the temperature change observed times the heat capacity of the calorimeter and its contents.

$$\Delta H = \Delta T \text{ (Heat capacity of calorimeter + Heat capacity of contents)}$$

where ΔT is the positive difference between the initial temperature and the final temperature.

We must determine the heat capacity of the calorimeter by first determining the amount of heat needed to raise its temperature by 1°C. This is accomplished by calculating the temperature change which occurs when a known amount of hot water is added to a known amount of cold water *in the calorimeter.* The specific heat of water is 4.18 joules per gram per degree. The heat lost by the warm water is used to heat the cold water and the calorimeter.

If T_1 represents the initial temperature of the calorimeter and 100 g of the cooler water, T_2 represents the initial temperature of 100 g of warm water to be added to the calorimeter and T_f the final temperature of the whole system after the hot and cold water are combined, then the heat lost by the hot water is

$$\text{Heat lost} = (T_2 - T_f) \text{ deg } (100 \text{ g}) \left(\frac{4.18 \text{ j}}{\text{g deg}} \right)$$

and the heat gained by the calorimeter and the cooler water is

$$\text{Heat gained} = (T_f - T_1)\deg \left[100\text{ g}\left(\frac{4.18\text{ j}}{\text{g deg}}\right) + \text{Heat capacity of calorimeter}\right]$$

If the calorimeter is well insulated, we can consider it to be an isolated system which does not lose heat to its surroundings. In this case the law of conservation of energy allows us to equate the heat lost by the hot water to the heat gained by the calorimeter and cool water:

$$(T_2 - T_f)(100)\text{ joules} = (T_f - T_1)(100 + \text{Heat capacity of calorimeter})$$

If we know T_1, T_2, and T_f (all determined experimentally), we can determine the heat capacity of the calorimeter.

Once the heat capacity of the calorimeter is known, it can be used to determine the heat changes in chemical reactions by carrying out the reactions inside the calorimeter. The heat of neutralization is an example of a reaction which can be studied in this way. The reaction of aqueous hydrochloric acid and aqueous sodium hydroxide proceeds as

$$Na^+ + OH^- + Cl^- + H^+ \rightarrow Na^+ + Cl^- + H_2O$$

The Na^+ and Cl^- ions are not directly involved in the reaction; they are *spectator ions*. The net reaction is the formation of water.

If this reaction is carried out in a calorimeter, we can determine the heat effect which accompanies it. The heat is due to the reaction

$$H^+ + OH^- \rightarrow H_2O$$

From the reaction of a known amount of acid with a known amount of base, we can determine the heat effect produced by the formation of a known number of moles of water.

The temperature of any object tends to approach the temperature of its environment. This means that a calorimeter which has been heated (by a chemical reaction) to a temperature above its surroundings will slowly lose heat and return to the temperature of its surroundings (Fig. 15-2). This is why we try to determine the maximum temperature attained in the calorimeter.

PROCEDURE

Numbers in parentheses refer to entry numbers on the data sheet.

I. DETERMINATION OF THE HEAT CAPACITY OF THE CALORIMETER Assemble the calorimeter as shown in Figure 14-1. Prop up the thermometer so it will not fall. Weigh the apparatus (1). Place 100 mL of water (measured with a graduated cylinder) into the calorimeter cup, and replace the cap and thermometer. Allow the system to equilibrate for 5 to 10 minutes and record the temperature as accurately as you can read the thermometer (2). While waiting for temperature equilibration, weigh the apparatus plus water (3).

In a separate beaker heat exactly 100 mL of water to about 30°C above room temperature. Use a Bunsen burner with a low flame to heat the water. When the water in the beaker has reached the desired temperature, remove the beaker from the heat and allow it to stand for a few minutes to reach a uniform temperature. Record the temperature of this water (4).

After you have recorded the temperature of the warm water, pour it carefully and completely into the calorimeter. Replace the calorimeter cover and swirl the flask gently to mix the contents. Record the temperature every 20 seconds until it is constant or drifts downward by a small constant between readings. Record the temperatures on your data sheet (5). Weigh the apparatus plus all added water again (7). The heat given up by the warm water as it cools is the product of its specific heat times the change in temperature times the weight of the water (9).

$$\text{Heat} = (\text{g water})\left(\frac{4.18\text{ j}}{\text{g deg}}\right)(T_i - T_f)\deg$$

The heat absorbed by the cool water is similarly calculated (11).

$$\text{Heat} = (\text{g water})\left(\frac{4.18 \text{ j}}{\text{g deg}}\right)(T_f - T_i) \text{ deg}$$

The heat lost by the hot water will be greater than the heat gained by the cool water, the difference going to the calorimeter (if we have a good calorimeter that will hold heat) (12). Now it is possible to calculate the heat capacity of the calorimeter. Divide the calories absorbed by the calorimeter by its temperature change (the same as the cool water) (13).

II. DETERMINATION OF THE HEAT OF REACTION: THE NEUTRALIZATION OF HYDROCHLORIC ACID BY SODIUM HYDROXIDE

Measure out exactly 100 mL of 3.0 M hydrochloric acid,* transfer it as completely as possible to the calorimeter, and record its temperature (14). Then measure out exactly 100 mL of 3.0 M sodium hydroxide.* Transfer it as completely as you can to the calorimeter. Close the calorimeter lid, stir the reacting solutions gently, and record the temperature every 20 seconds on your data sheet until a constant temperature is reached or the temperature drifts downward by a small constant amount between readings (15). Record the maximum temperature (16) and proceed with the calculations (17–20) as before. The number of moles of water produced is 0.3 mole (3 moles/liter × 0.1 liter) (20). Finally, the joules per mole of water produced can be calculated (21).

In the calculations we will assume that the heat capacity of the water used in Part I and the solution produced in Part II is 4.18 joules per gram per degree. This approximation is well within the limits of measurement imposed by the apparatus used in this experiment.

* The solutions of hydrochloric acid and sodium hydroxide should be prepared well ahead of the time of their use so there is adequate opportunity for them to equilibrate to the same room temperature.

PRE-LAB QUESTIONS

Heat of Neutralization Experiment

1. What specific chemicals in this experiment could burn your skin or harm your eyesight if proper safety precautions aren't observed? Give formula(s).

   _____

2. Is the heat of neutralization reaction endothermic or exothermic?

   _____

3. A. Write the a complete balanced chemical equation for the reaction of aqueous hydrochloric acid with aqueous sodium hydroxide.

   _____

   B. Write the net ionic equation for the reaction given in (A).

   _____

4. The amounts and concentrations of HCl(aq) and NaOH(aq) given in the procedure section produce 0.3 moles of water when they are mixed. Show with calculations the basis for this number of moles.

Experiment 14
Data Sheet

Name _____

Date _____

CALORIMETRY

DETERMINATION OF THE HEAT CAPACITY OF THE CALORIMETER

(1) Weight of empty calorimeter and thermometer _____ g

(2) Temperature of calorimeter and water before mixing _____ °C

(3) Weight of calorimeter + thermometer + cool water _____ g

(4) Temperature of warm water _____ °C

(5) Temperature after mixing:

   20 sec _____ °C   40 sec _____ °C   60 sec _____ °C

   80 sec _____ °C   100 sec _____ °C   120 sec _____ °C

(6) Temperature maximum after mixing (T_f) _____ °C

(7) Weight of calorimeter + thermometer + all water _____ g

(8) Weight of warm water (subtract 3 from 7) _____ g

(9) Heat furnished by cooling of warm water

   (wt. of warm water $\times$ temperature increase $\times$ 4.18 j/g deg) _____ j

(10) Weight of cool water (subtract 1 from 3) _____ g

(11) Heat absorbed by cooler water

   (wt. of cool water $\times$ temperature increase $\times$ 4.18 j/g deg) _____ j

(12) Heat absorbed by calorimeter (subtract 11 from 9) _____ j

(13) Heat capacity of calorimeter

   (divide 12 by the temperature increase) _____ j/deg

DETERMINATION OF THE HEAT OF REACTION

(14) Temperature of calorimeter and acid before reaction _____ °C

(15) Temperature after adding sodium hydroxide solution to the calorimeter:

   20 sec _____ °C   40 sec _____ °C   60 sec _____ °C

   80 sec _____ °C   100 sec _____ °C   120 sec _____ °C

(16) Temperature maximum after mixing _____ °C

(17) Observed temperature change (subtract 14 from 16) _____ °C

(18) Joules added to calorimeter during the reaction

   = ΔT (Heat capacity of calorimeter) _____ j

(19) Joules added to solution during reaction

     $= \Delta T \times$ Mass of solution $\times$ 4.18 j/g deg   _____ j

(20) Total joules released by reaction (add 18 and 19)   _____ j

(21) Moles of water produced during reaction   _____

(22) Joules released per mole of water (divide 20 by 21)   _____ j

POST-LAB QUESTIONS

Heat of Neutralization Experiment

1. In carrying out the heat of neutralization experiment, a student added 90.0 mL instead of 100.0 mL of 3 M HCl to 100.0 mL of 3 M NaOH. Would the calculated value for the heat of neutralization be (high, low, unaffected) by this procedural error? Explain.

    _____

    _____

    _____

2. In carrying out the heat of neutralization experiment, the temperature of 3 M hydrochloric acid was lower than the temperature of 3 M sodium hydroxide before mixing. Would the calculated value for the heat of neutralization be (high, low, unaffected) by this procedural error? Explain.

    _____

    _____

    _____

3. A. A student mixed exactly 100.0 mL of 1.000 M sodium hydroxide with 100.0 mL of 1.000 M hydrochloric acid in a calorimeter. The initial temperature of both solutions was 23.5°C and the highest temperature reached by the mixture was 30.7°C. Assume the heat capacity of the calorimeter is zero. How many joules of heat were produced?

    _____

    _____

    _____

    B. How many moles of water were formed in (A)?

    _____

    _____

    _____

    C. Calculate the heat of neutralization in kj for one mole of water using your results from (A) and (B).

    _____

    _____

    _____

15 Titration of an Acid with a Base

OBJECTIVE

To become familiar with a precise analytical technique.
To determine the amount of acid in a solution by titration.

APPARATUS AND CHEMICALS

50-mL buret (2)
buret clamp
ring stand
250-mL flask (3)
2 small funnels

250-mL beakers (2)
phenolphthalein solution
standard sodium hydroxide solution
unknown solution of hydrochloric acid

SAFETY CONSIDERATIONS

Use care in handling all glassware to prevent breakage. The burets can be broken at the clamp if excess pressure is applied.

The acid and base used in this experiment are dilute enough that they do not pose a serious hazard to skin and clothes if washed off quickly. However, either can cause serious and immediate damage to the sensitive tissues of the eyes. Safety glasses must be worn at all times, and if acid or base does get into the eyes, the eyes must be flushed immediately for 15 minutes with cold water and the instructor notified at once.

FACTS TO KNOW

The word *titration* is derived from a French word, *titre*, which means "to bestow a title upon or to standardize." The purpose of a chemical titration is to standardize a solution, that is, to determine the concentration of a substance in a solution. The *concentration* of a substance in a solution is the *amount* of the substance dissolved in a certain *volume* of solution. Often, concentrations of solutions are given in terms of molarity, M, which is defined as the number of gram molecular weights (or moles) in a liter of solution. For example, the molecular weight of HCl is 36.5. If 73 g (2×36.5) are dissolved in sufficient water to make a liter of solution, the concentration of the HCl in the solution is 2 molar (2 M), since 2 molecular weights in grams (or moles) are dissolved in the liter of solution.

In this experiment, the concentration of an acid in a solution will be determined by measuring the amount of acid solution required to react completely with a known amount of base. The acid will be hydrochloric acid, an important component of the digestive juices of the stomach, and the base will be sodium hydroxide, commonly known as lye. The equation for the reaction between hydrochloric acid and sodium hydroxide can be written

$$HCl + NaOH \rightarrow H_2O + NaCl$$

In order to determine the amount of hydrochloric acid in a solution, we must first measure a precise volume of this unknown acid. The second step requires that we add just enough sodium hydroxide solution of known concentration (standard solution) to neutralize the acid. This standard solution of sodium hydroxide has been carefully prepared to contain a rather exact amount of sodium hydroxide per liter of solution. Since we know the amount of sodium hydroxide per liter and the volume required to neutralize the measured volume of unknown hydrochloric acid concentration, we can calculate the amount of sodium hydroxide needed by multiplying the concentration of the standard base by the measured volume required for neutralization of the acid. For example,

$$amount/volume \times volume = amount,$$

or

$$concentration \times volume = amount.$$

When sufficient base solution has been added to neutralize completely the amount of acid present, a visual indicator changes color to signal that all of the base has reacted. In this experiment, the indicator, phenolphthalein, is pink in basic solution and colorless in acidic solution. Hence, when the solution is a very faint pink, sufficient sodium hydroxide has been added to the solution of hydrochloric acid to just neutralize the HCl and have a very slight amount of NaOH left over.

Volumes of solutions can be measured easily to an accuracy of ±0.02 mL by using burets like those shown in Figure 15-1. To read a buret, have your eye level with the *bottom* of the curved surface of the liquid (the meniscus). To check the level of your eye, use the rings marked on the buret. If the ring nearest the meniscus appears to be a straight line, then your eye is at the proper height to read the buret. Now, place a contrast card behind the buret, as shown in Figure 15-2. The black portion is held just below the bottom of the meniscus, so that the meniscus appears to be black against a white background. Read the volume to two decimal places. The second decimal place (the hundredths place) is estimated by mentally dividing the space between two lines into ten equal parts, and counting the number of these parts required to go from the line just above the meniscus to the bottom of the meniscus.

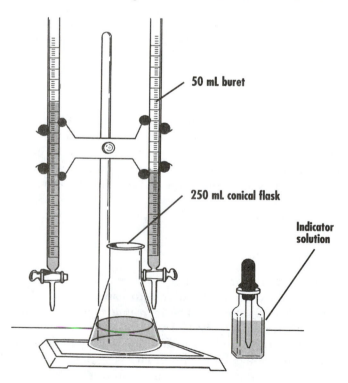

Figure 15-1. Equipment for titration.

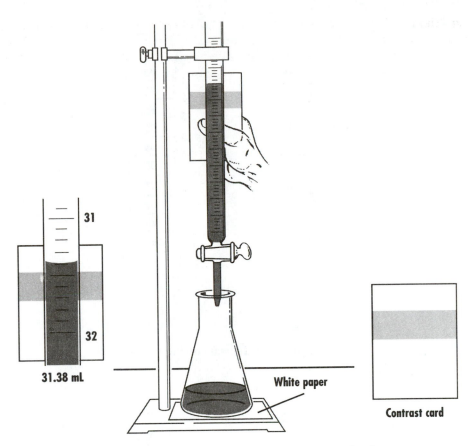

31

32

31.38 mL

White paper

Contrast card

Figure 15-2. Proper use of a contrast card to read volumes.

To gain a clear understanding of the process of a titration and the reasons for the calculations you will make, consider the following simplified titration.

Suppose you know that 1 atom (or ion) of acid A reacts with 1 atom (or ion) of base B to form a molecule of AB. That is:

$$A + B \rightarrow AB$$

Now suppose that we have a solution of A containing 3 A's per mL and a solution of B containing 2 B's per mL as shown in Figure 15-3.

Begin the titration by running in 3 mL of A and then add an indicator (Fig. 15-4). At this point we have 9 A's to be neutralized by the base; that is:

$$\text{amount of A} = (\text{concentration of A}) \times (\text{volume of A})$$

$$= \frac{3 \text{ A's}}{\text{mL}} \times 3 \text{ mL}$$

$$= 9 \text{ A's}$$

Now add solution containing B until you have just as many B's in solution as you have A's, that is, until you have only AB's in solution, with no excess of either A or B. If a very slight excess of B is added, the indicator changes color.

How many mL of B do we have to add? We have to add enough mL of B to give 9 B's, to match the 9 A's previously added. This means that we have to add 4.5 mL of B (Fig. 15-5). At the endpoint,

$$\text{amount of A} = \text{amount of B}$$

$$9 \text{ A's} = \frac{2B}{\text{mL}} \times 4.5 \text{ mL}$$

$$9 \text{ A's} = 9 \text{ B's}$$

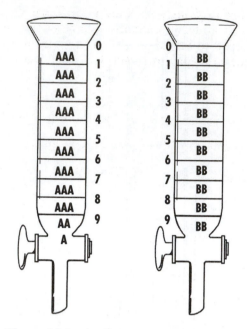

Figure 15-3. At the start of the titration.

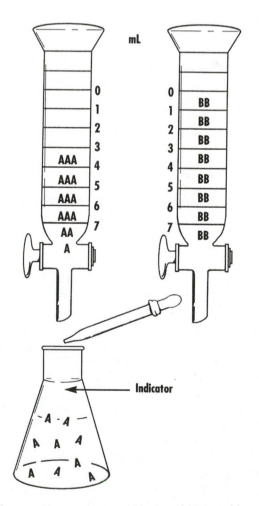

Figure 15-4. After A's have added to flask.

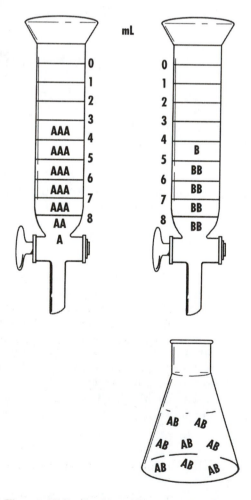

Figure 15-5. At the end of the titration.

In this example, both concentrations were known, but normally in a titration only one concentration is known, along with the volume of A and the volume of B required to reach an endpoint. The volumes can always be determined in the laboratory. The basis for calculating the concentration of a substance of unknown concentration (the hydrochloric acid, in this experiment) is exactly the same as in the example.

At the endpoint,

$$amount\ of\ acid = amount\ of\ base$$

$$concentration\ of\ acid \times volume\ of\ acid = concentration\ of\ base \times volume\ of\ base$$

$$concentration\ of\ acid = \frac{concentration\ of\ base \times mL\ base}{mL\ acid}$$

PROCEDURE

Numbers in parentheses refer to entry numbers on the data sheets.

I. The sodium hydroxide solution has been standardized. Write its concentration as given on the bottle on the data sheet (1). Obtain about 150 mL of this standard sodium hydroxide solution in a *clean, dry* beaker.

II. Set up two clean burets as shown in Figure 16-1. Clean each buret until drops of water do not cling to the inside as the buret is drained (Fig. 16-6).

It is neither necessary nor desirable to dry the buret after cleaning.

Figure 15-6. Water drops should not cling to buret.

III. Rinse each buret with two portions of its specific solution in the following manner: Remove the right-hand buret from the assembly and add about 5 mL of the standard sodium hydroxide solution to the buret. A dry funnel may be used or you may pour directly from the beaker. Place your thumb over the open end of the buret and invert the buret several times in order to rinse the inside surface. Under running water, rinse the solution off your fingers. Drain this solution through the tip into a beaker and discard this portion. Finally, fill the buret slightly above the zero mark with sodium hydroxide solution and clamp the buret on the ring stand. Be sure the buret is vertical.

IV. To remove the air bubbles from the tip of the buret, open the stopcock and allow rapid solution flow to flush out air bubbles.

V. With a clean, dry beaker, obtain approximately 150 mL of the unknown acid solution from the instructor. Remove the left-hand buret from the assembly, rinse it twice with 5-mL portions of the acid solution, fill it with acid solution to slightly above the zero mark, remove the air bubbles, and wash any acid off of your hands.

VI. Drain both burets to slightly below the zero mark, then read and record to *two decimal places* the liquid level in each buret (2 and 3). Be careful to follow the reading technique as described in the introduction.

VII. Run out approximately 25 mL of the acid solution into a clean flask. Do not read the exact volume of acid at this time; wait until the titration is completed, because more acid may be required.

Add 4 drops of the phenolphthalein solution to the flask.

Carefully place the flask under the buret containing base and slowly add base, with swirling (Fig. 16-7), until the indicator just changes to the slightest tinge of pink. If too much base is added, add more acid solution until the pink color just disappears. Then add one-half drop of the base, or more if necessary, to restore the slightest tinge of pink. **PATIENCE!**

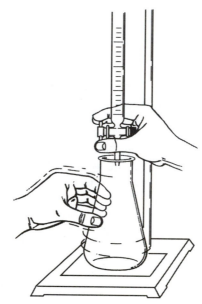

Figure 15-7. Swirl flask during titration.

When a satisfactory endpoint is reached, record the final readings on the burets to two decimal places (4 and 5). Obtain the volumes of acid and base used by difference (6 and 7).

Repeat the titration by refilling the burets with the appropriate solutions. Rinsing is unnecessary the second time.

VII. Calculate the concentration of the acidic solution for each titration (8).

$$\text{conc. of acid} = \frac{\text{conc. of base} \times \text{mL of the base}}{\text{mL of acid}}$$

If the concentrations calculated from these two titrations differ by more than 0.005, a third titration must be performed.

Finally, take the average of the concentrations of acid by adding up all of the concentrations and dividing by the number of titrations performed. Report the average concentration of acid solution to two decimal places (9).

EXAMPLE CALCULATION

Suppose 20.25 mL NaOH is required to neutralize 10.13 mL HCl and the concentration of NaOH is 0.105 M. Find the concentration of HCl:

$$\text{concentration of HCl} = 0.105 \times \frac{20.25}{10.13} = 0.210 \text{ M}$$

PRE-LAB QUESTIONS

1. What would be the effect of placing the standard sodium hydroxide in a beaker which contained some residual water in the beaker?

    _____

    _____

    _____

2. If you were to splash a basic solution into your eyes, what should be your immediate response?

    _____

    _____

    _____

3. If one had one gallon of sodium hydroxide solution which had a concentration of two pounds of sodium hydroxide per gallon, what would be the concentration of four gallons of that acid?

    _____

    _____

    _____

4. Suppose you had 5 gallons of sodium hydroxide solution with a concentration of two pounds of sodium hydroxide per gallon. What is the total amount of sodium hydroxide?

    _____

    _____

    _____

5. What is the purpose of the phenolphthalein solution?

    _____

    _____

    _____

Experiment 15
Data Sheet

Name _____

Date _____

TITRATION OF AN ACID WITH A BASE

(1) Concentration of the standard base   _____ molar

	Run 1	Run 2	Run 3	Run 4
(2) Acid initial reading	_____	_____	_____	_____
(3) Base initial reading	_____	_____	_____	_____
(4) Acid final reading	_____ mL	_____ mL	_____ mL	_____ mL
(5) Base final reading	_____ mL	_____ mL	_____ mL	_____ mL
(6) Total mL used (acid)	_____	_____	_____	_____
(7) Total mL used (base)	_____	_____	_____	_____
(8) Concentration of unknown acid	_____ M	_____ M	_____ M	_____ M
(9) Average concentration of unknown acid				_____ M

Do your calculations in an organized fashion below:

POST-LAB QUESTIONS

1. How do you know when you have added too much standard base (exceeded the endpoint)?

_____

_____

_____

2. Why is it unnecessary to measure precisely 25.00 ml of unknown acid into the beaker?

_____

_____

_____

3. If all the students placed their "unused" standard sodium hydroxide back into the original container what would be the likely effect on the next class?

_____

_____

_____

4. How much 0.150 M sodium hydroxide would be required to just neutralize 15 ml of 0.175 M hydrochloric acid?

_____

_____

_____

16 Standardization of a Sodium Hydroxide Solution

OBJECTIVE

To use a primary standard subject substance to prepare a sodium hydroxide solution of accurately known molarity.

APPARATUS AND CHEMICALS

Vial of reagent grade potassium acid phthalate (3.5 g)
$3M$ NaOH (34 mL)
50 mL buret and buret clamp
250 mL Erlenmeyer flask (3)
600 mL beaker (1)
laboratory balance (± 0.01 g)

deionized water
drying oven
desiccator
phenolphthalein indicator
analytical balance (± 0.0001 g)

SAFETY CONSIDERATIONS

3 M NaOH is a corrosive liquid. Wear your safety goggles at all times.

FACTS TO KNOW

Acid-base titrations are an important analytical method for a chemist. The exact quantity of acid present in a certain mixture can be found by determining the quantity of base required to just neutralize the acid. Thus, a titration is the process by which a solution containing a known concentration of a substance (the titrant) is added to another solution containing an unknown concentration of a second material that will react with the titrant. For any titration it is necessary to have (1) a solution of known concentration, (2) a means of detecting the equivalence point, i.e., when the reaction is complete, and (3) calibrated glassware such as burets, pipets, and volumetric flasks.

The net ionic equation for acid-base titrations is the neutralization reaction

$$H_3O^+ + OH^- \rightarrow 2\,H_2O$$

Thus 1 mole of H_3O^+ reacts with one mole of OH^-. An indicator is often used to detect when exactly equivalent numbers of moles of acid and base have been added together. The indicators used in acid-base titrations are weak organic acids or bases that change color when their degree of protonation is changed. One of the most common indicators is phenolphthalein, which is colorless in acid solutions but becomes pink when the solution becomes basic.

123

In the present experiment you will standardize a solution of NaOH by using the *primary standard* potassium hydrogen phthalate. Solutions of acids and bases are commonly standardized by taking a solution with approximately the desired concentration and titrating against an accurately weighed quantity of a pure compound. Potassium hydrogen phthalate. $KHC_8H_4O_4$, is a monobasic acid which contains 1 mole of neutralizable hydrogen ions per mole of compound. Its molar mass is 204.22 g. It is pure, non-hygroscopic, thermally stable compound (at 110°C) which can be weighed out to provide an exact number of moles of titratable H^+. Sodium hydroxide does not have such characteristics. It absorbs moisture and CO_2 quickly from the air. As a result, the actual base concentration of a NaOH solution is found by standardization.

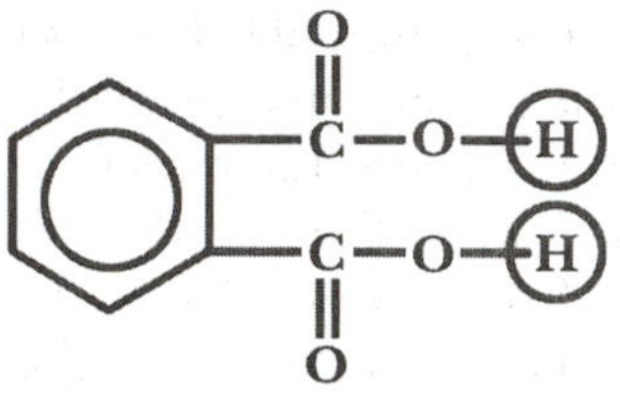

Phthalic acid has two ionizable hydrogen ions. The primary standard used in this experiment is produced by reacting one mole of potassium hydroxide with one mole of phthalic acid to produce potassium hydrogen phthalate which still has one ionizable hydrogen ion.

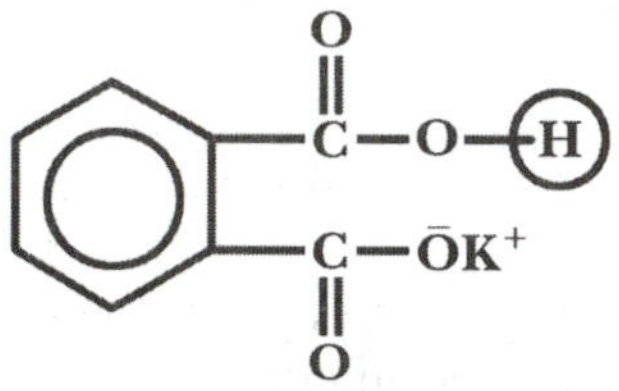

Ionizable hydrogen ion in Potassium hydrogen phthalate

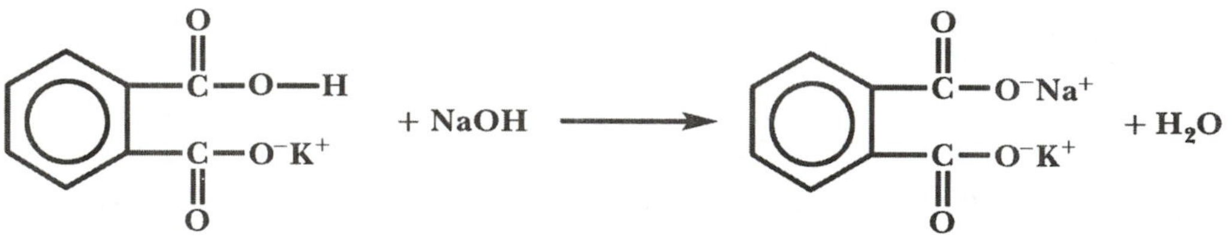

Figure 16-1

PROCEDURE

I. Make sure your buret is clean. If not, wash the buret thoroughly with soap and water, rinse first with tap water, then 3-4 times with 5mL portions of deionized water. When the buret is clean, water should drain without leaving droplets on the inside surface. When you finish your lab work, rinse the buret with deionized water.

II. Add 200 mL of deionized water to a 600 mL beaker. Then add 34 mL of 3 M NaOH and 266 mL more of dionized water. Mix well with a stirring rod for 5 minutes. This NaOH will next be standardized and should be saved for subsequent use in the titration of commercial products experiment

III. Remove one of the vials of potassium hydrogen phthalate (KHP) from the oven. Cool the vial and its contents in a desiccator. Weigh the KHP vial on the analytical balance. Record the weight (1). Tare a 125 mL Erlenmeyer flask on a laboratory balance and transfer about 0.9 - 1.0 g of the KHP from the vial into a 125 mL Erlenmeyer flask. Then reweigh the vial of KHP on the analytical balance and record the weight (2).

IV. Add about 40 mL of deionized water to the flask to dissolve the KHP. Mark this flask #1.

V. Prepare another sample of KHP following the same procedure and mark this flask #2.

VI. Set up the buret in a buret clamp on a ring stand. With the stopcock closed add about 20 mL of your NaOH solution to rinse the buret. Drain through the stopcock. Then fill the buret above the zero mark with your NaOH solution and then open the stopcock to allow solution to run into the tip of the buret. Make sure there are no air bubbles in the tip. Adjust the level of the solution at or below the zero mark. Record an initial reading of the buret (two decimal places) (5).

VII. Add 3 drops of phenolphthalein indicator to both flask #1 and flask #2. Place #1 under the buret and slowly add NaOH, with swirling, until the indicator just changes to the slightest tinge of pink. Remember that the base should be added slowly near the endpoint. When it appears that the endpoint is near, i.e., the point at which the entire solution just becomes permanently pink, the base should be added so slowly that the color disappears between the addition of each drop and the next. Continue to add base until the pink color barely persists throughout the whole solution for at least 40 seconds. Record the final buret reading (4).

VIII. Repeat the titration with flask #2 by recording an initial buret reading, titrating to a permanent pale pink color, and recording the final buret reading.

IX. Calculate the molarity of the NaOH solution for each sample (7)–(9). If the concentrations calculated from these two titrations differ by more than 0.005, do a third titration. Finally, take the average of the concentrations of NaOH, and record it to three significant figures. Put the remaining NaOH solution in a *dry* 250 mL Erlenmeyer flask, place a label on it with the molarity, cover with Saran wrap, and save for the next experiment.

PRE-LAB QUESTIONS

1. Calculate the molar mass of potassium hydrogen phthalate. Does your value agree with that given in the experiment?

    _____

    _____

    _____

    _____

2. In the calculation section, the statement is made that potassium hydrogen phthalate combines with sodium hydroxide in a 1:1 mole ratio. What is the basis for this statement?

    _____

    _____

    _____

    _____

3. Why is it necessary to standardize sodium hydroxide solution, rather than simply weighing out an accurate amount of solid sodium hydroxide and dissolving it in a known volume of water?

    _____

    _____

    _____

    _____

4. Why is it recommended to titrate to a pale pink color at the endpoint rather than a deep pink color?

    _____

    _____

    _____

    _____

Experiment 16   **Name** _____

Data Sheet   **Date** _____

STANDARDIZATION OF A SODIUM HYDROXIDE SOLUTION

	Sample 1	**Sample 2**	**Sample 3**
(1) Initial weight of vial of KHP	_____	_____	_____
(2) Final weight of vial of KHP	_____	_____	_____
(3) Weight of potassium hydrogen phthalate	_____	_____	_____
(4) NaOH burette, final reading	_____	_____	_____
(5) NaOH burette, initial reading	_____	_____	_____
(6) Volume of NaOH used	_____	_____	_____

Calculations: Since potassium hydrogen phthalate and NaOH combine in a 1:1 mole ratio, moles potassium hydrogen phthalate = moles NAOH so

$$\text{Molarity of NaOH} = \frac{\text{moles of potassium hydrogen phthalate}}{\text{liters NaOH used}}$$

(7) First sample: (8) Second sample:

(9) Third sample if needed:

Record molarity for each sample here: _____ _____ _____

                           Average: _____

POST-LAB QUESTIONS

1. A 0.4168 g sample of potassium acid phthalate was dissolved in 50 mL of deionized water and titrated to a phenophthalein end-point with 19.76 mL of NaOH solution. What is the molarity of the NaOH?

    _____

    _____

    _____

    _____

2. Suppose a student used 60 mL of deionized water to dissolve the potassium hydrogen phthalate sample instead of the recommended 40 mL. Would this affect the results of the titration with NaOH solution? Explain your answer.

    _____

    _____

    _____

    _____

3. A student failed to remove air bubbles in the buret tip before starting the titration. About halfway through the titration, the air bubble disappeared and the tip was full of NaOH solution. Would the resulting calculation of the NaOH molarity be too low or too high? Explain your answer.

    _____

    _____

    _____

    _____

Titration of Commercial Products

OBJECTIVE

To determine the concentration of acetic acid in vinegar and the amount of magnesium hydroxide in milk of magnesia tablets.

APPARATUS AND CHEMICALS

standard 0.200 M NaOH (student prep)   10-mL graduated pipet
standard 1.000 M NaOH   pipet bulb
50-mL buret (2)   white vinegar
standard 1.000 M HCl   milk of magnesia tablets
125-mL Erlenmeyer flask (2)   analytical balance
250-mL Erlenmeyer flask (2)   Phenolphthalein

SAFETY CONSIDERATIONS

The acidic and basic solutions used in this experiment can cause skin and eye damage. Wear your safety goggles at all times. Always use a pipet bulb for pipetting solutions.

FACTS TO KNOW

Several titrations of commercial products will be carried out. These include the titration of vinegar with standardized sodium hydroxide solution and the titration of milk of magnesia tablets with standardized hydrochloric acid solution.

✳ Vinegar is an aqueous solution of acetic acid. Acetic acid is an organic acid with the formula CH_3COOH. The titration reaction is:

$$| \quad CH_3COOH(aq) + |NaOH(aq) \rightarrow Na^+CH_3COO^-(aq)\ 1\ H_2O(l)$$
ᴬᶜᵉᵗᶦᶜ ᴬᶜᶦᵈ   ᴮᴬˢᴱ

In this experiment the amount of acetic acid in vinegar will be checked against the *federal requirement of a minimum of 4 g of acetic acid per 100 mL of vinegar.*

Magnesium hydroxide is the active ingredient of milk of magnesia tablets that are used to neutralize excess acidity in the stomach. The titration reaction is

$$2HCl(aq) + Mg(OH)_2(s) \rightarrow MgCl_2(aq) + 2\ H_2O(l)$$

Magnesium hydroxide has a limited solubility in water. An equilibrium exists between solid magnesium hydroxide and dissolved magnesium ions and hydroxide ions.

$$Mg(OH)_2(s) \Leftrightarrow Mg^{2+}(aq) + 2\ OH^-(aq)$$

All the magnesium hydroxide can be dissolved by adding a reagent that will shift the equilibrium to the right. Adding excess hydrochloric acid solution will shift the equilibrium to the right until all $Mg(OH)_2$ is dissolved because the H^+ ions from $HCl_{(aq)}$ neutralize the OH^- ions from magnesium hydroxide to form H_2O. In the present experiment an excess of 1.000 M HCl(aq) will be added, and the excess HCl above that needed to neutralize magnesium hydroxide will be back-titrated with standardized NaOH(aq). This will provide information on the exact amount of acid needed to neutralize the magnesium hydroxide in a milk of magnesia tablet.

Milk of magnesia tablets also have a binder (usually starch) which holds tablets together. The binder usually is not water soluble. As a result, the solution still appears cloudy after $Mg(OH)_2$ has dissolved.

PROCEDURE

A. Acetic Acid in Vinegar

You will be using the standardized NaOH you prepared in the previous experiment. Record the concentration of your standardized NaOH on the data sheet (6) and (11).

I. Make sure your buret is clean. If not, wash the buret thoroughly with soap and water, rinse first with tap water, then 3–4 times with 5 mL portions of deionized water. When the buret is clean, water should drain without leaving droplets on the inside surface. When you finish your lab work, rinse the buret with deionized water.

II. Rinse your buret twice with 10 mL portions of your standardized NaOH solution from the previous experiment. Then fill a buret, then drain to zero or slightly below, making sure there are no air bubble in the tip.

III. Use a 10 mL graduated pipet and a pipet bulb to dispense 5 mL of vinegar samples into each Erlenmeyer flask (1). Label these flasks #1 and #2. Wash down the sides of these flasks with deionized water.

IV. Record the initial reading on the buret (3) and add ~~two~~ 5 drops of phenolphthalein to flask #1. Titrate to a pale pink endpoint and record the final buret reading (2).

V. Repeat procedure IV for flask #2, being sure to add phenolphthalein and recording initial and final buret readings (3) and (2).

B. Titration of Phillips Milk of Magnesia

Hint: If the tablet samples of milk of magnesia are prepared early in the lab period, they should be completely softened when you are ready to titrate them..

I. Weigh a tablet accurately using a weighing boat and the analytical balance (1) and (2). Transfer the tablet to a 250 mL Erlenmeyer flask that contains 150 ml of deionized water. Label this flask #1. Allow the tablet to soften for about 10 min. Carefully crush the tablet with a stirring rod, rinsing the rod into the flask when it is removed.

II. Prepare a second sample by following procedure in I for a second tablet (1) and (2). Label this flask #2.

III. Add two drops of phenolphthalein to both flasks. At this point you are ready to add a known amount of excess 1.000 M HCl. Several students can share the same HCl buret. The buret should be rinsed with the 1.000 M HCl and then filled, taking care not to have bubbles of air in the tip. Take an initial reading on the HCl buret and record (5). Add between 12 and 13 mL of 1.000 M HCl and record the final reading (4).

IV. *Wait ten minutes* to allow time for the $Mg(OH_2)$ to dissolve as the OH^- in solution combines with the H^+ from the hydrochloric acid. (Review discussion of equilibrium in Facts To Know section.) While you are waiting, refill the standardized NaOH solution. The back-titration to a faint-pink endpoint will take you from about 8 mL to as much as 15mL. this depends on how much HCl you added and the concentration of your standardized NaOH. Record the final reading (8) if the endpoint is a good one (pale pink). If you added too much and missed the endpoint, take an initial reading on the HCl buret and add 0.40 mL of 1 M HCl and then add your standardized NaOH to a pale pink endpoint. Remember to record all readings so you know the volumes of HCl and NaOH you've added. Space is provided on the data sheet for additional readings. (12) - (18)

V. Repeat Procedure IV for flask #2.

PRELAB QUESTIONS

1. Write balance chemical equations for
   A. titration of acetic acid with sodium hydroxide

   B. titration of magnesium hydroxide with hydrochloric acid

2. A. How many moles of acetic acid, CH_3COOH, will combine with one mole of NaOH? _____

   B. How many moles of magnesium hydroxide will combine with one mole of HCl? _____

3. If $Mg(OH)_2$ is insoluble, why is it possible to carry out a titration reaction with standardized HCl(aq) to quantitatively measure the amount of OH^- present?

4. How many moles of NaOH are present in 500 mL of a 0.1 M NaOH solution?

5. A milk of magnesia tablet weighing 0.5850 g was found to contain 4.834×10^{-3} moles of $Mg(OH)_2$. What is the percentage of $Mg(OH)_2$ in the tablet?

Name _____

Date _____

TITRATION OF COMMERCIAL PRODUCTS

		#1	#2
A.	**Acetic Acid in Vinegar**		
(1)	Volume of vinegar in liters	.005	.005
(2)	Final NaOH buret reading		
(3)	Initial NaOH buret reading	.5	5
(4)	Volume of NaOH used		
(5)	Molarity of NaOH	1 m LITER	1M LITER

B. Titration of Milk of Magnesia Tablets

(1) Weight of tablet + weighing boat

(2) Weight of weighing boat

(3) Weight of tablet

(4) Final HCl buret reading

(5) Initial HCl buret reading

(6) Volume of HCl used

(7) Molarity of HCl used

(8) Final NaOH buret reading

(9) Initial NaOH buret reading

(10) Volume of NaOH used

(11) Molarity of NaOH used

This section is needed only if the endpoint was missed and additional 1.000 M HCl was added and back-titrated with your standardized NaOH.

(12) Final HCl buret reading

(13) Initial HCl buret reading

(14) Volume of HCl used

(15) Molarity of HCl used

(16) Final NaOH buret reading

(17) Initial NaOH buret reading

(18) Volume of NaOH used

Calculations

A. Acetic Acid in Vinegar

1. Moles of NaOH used: Trial 1:   Trial 2:

         Moles of NaOH used _____   _____

2. Molarity of acetic acid in vinegar: If you know how many moles of NaOH was used to titrate the acetic acid present in vinegar, how many moles are present in the 5 mL sample?

                                                Trial 1:   Trial 2:

         Moles of acetic acid present _____   _____

3. Molarity of acetic acid in vinegar:

                                                Trial 1:   Trial 2:

         Molarity of acetic acid in the vinegar _____   _____

4. Use the average molarity of your vinegar sample to calculate the mass of acetic acid per 100 mL of vinegar and compare this to the federal minimum requirement of 4 g/100 mL. Hint: M = moles/liter so moles = $(M)(V_{liters})$

B. Titration of Milk of Magnesia Tablets

1. *Moles of HCl Used* - Hint: Use moles HCl = $(HCl_{liters})(M_{HCl}) - (NaOH_{liters})(M_{NaOH})$

    Trial 1:   Trial 2:

    _____ moles HCl   _____ moles HCl

2. *Moles of Mg(OH)$_2$ Present* - Hint: Be sure to check the equation given for the titration reaction in the Facts To Know section.

Trial 1:   Trial 2:

_____ moles Mg(OH)$_2$ _____ moles Mg(OH)$_2$

3. *Grams of Mg(OH)$_2$ Present*

Trial 1:   Trial 2:

_____ g Mg(OH)$_2$ _____ g Mg(OH)$_2$

4. *Percentage Mg(OH)$_2$ in sample*

Trial 1:   Trial 2:

_____ % Mg(OH)$_2$ _____ % Mg(OH)$_2$

Average % Mg(OH)$_2$ _____

POSTLAB QUESTIONS

1. In analyzing milk of magnesia tablets why is it necessary to add excess 1.000 M HCl and back-titrate with standardized NaOH rather than just titrating with 1.000 M HCl to the point where the pink solution turns clear?

2. Suppose a student added excess 1.000 m HCl and immediately started back-titrating with standardized NaOH rather than waiting ten minutes. Would the calculated amount of $Mg(OH)_2$ be too high or too low? Explain.

3. Calculate the percentage $Mg(OH)_2$ contained in a milk of magnesia tablet from the following data. Given: 13.00 mL of 1.000 M HCl was added to a 0.5800 g tablet of milk of magnesia. The solution was back-titrated to a phenolphthalein endpoint with 12.70 mL of 0.2040 M NaOH.

4. A. Calculate the molarity of acetic acid in vinegar for the following set of data. Given: 5.00 mL of vinegar required 16.90 mL of 0.1992 M NaOH to reach the phenolphthalein endpoint.

    B. Calculate the grams/100 mL of acetic acid in the vinegar sample in part (A) to see if it meets the federal minimum requirement of 4 g/100 mL.

EXPERIMENT 18

Qualitative Analysis

OBJECTIVE

To demonstrate the procedures by which unknown substances in solution may be isolated and identified by making use of their unique properties.

APPARATUS AND CHEMICALS

Bunsen burner
ring stand
iron ring
plastic funnel
250 mL beaker
plastic wash bottle
red & blue litmus paper
filter paper
6 small test tubes in rack

0.2 M $AgNO_3$
O.2 M $Pb(NO_3)_2$
0.2 M $Hg_2(NO_3)_2$
0.2 M $Cu(NO_3)_2$
0.2 M $SnCl_4$
0.2 M $Bi(NO_3)_3$
1.0 M K_2CrO_4
0.1 M $SnCl_2$
1.0 M CH_3CSNH_2 (thioacetamide)

6 M HCl
6 M NH_4OH
6 M HNO_3
0.1 M NH_4Cl
3 M KOH
6 M $HC_2H_3O_2$
1 M $K_4Fe(CN)_6$

SAFETY CONSIDERATIONS

 The solutions you are using are either very corrosive acids or bases, or metal ion solutions, many of which are toxic. Treat all solutions with respect and dispose of them only as directed by your instructor. The acids and bases are very damaging to soft tissues (eyes) and to clothes. Wear safety glasses at all times and wear old clothes or protective aprons.

 Observe all precautions regarding hair and clothing when using the bunsen burner.

Be careful with the glassware. The test tubes are often hot and are sometimes difficult to hold.

FACTS TO KNOW

Chemical analysis is the examination of a material in order to determine either (a) its identity or the identity of a component of the material or (b) the exact makeup of a substance or the exact percentage of a component of the substance. One might examine a sample of material found in an old mine to determine if it was an ore which contained uranium, copper, iron, or other valuable mineral.

Upon determining that the ore contained iron, one might then examine it further to find the exact percentage of iron in the ore to determine the economic feasibility of mining it. The examination to determine what is present in the ore is an example of *qualitative analysis*, whereas the examination of the ore known to contain iron to determine the exact percent of iron present is an example of *quantitative analysis*. The very nature of the analytical procedure to be used for a particular analysis depends heavily on which of these two goals directs the analyst.

Quantitative analysis (as the prefix "quantit(y)" suggests) is intimately concerned with numbers. It is simplified somewhat by the prior knowledge of what is present, but is complicated by the necessity to be terribly careful with the sample, as any amount that is spilled or otherwise lost will render the analysis incorrect. Many of the techniques utilized in this manual involve quantitative analysis. *Gravimetric* quantitative analyses are done primarily by weighing starting materials and products (Experiment 11 - Determination of the Empirical Formula of a Compound; Experiment 12 - Determination of the Formula of a Hydrate; Experiment 13 - Gravimetric Determination of Sulfate). Volumetric quantitative analyses may be done by very carefully comparing volumes of solutions that react with one another (Experiment 15 - Titration of an Acid with a Base; Experiment 16 - Standardization of a Sodium Hydroxide Solution, Experiment 17 - Titration of Commercial Products; Experiment 23 - Analysis of Vitamin C). Many quantitative analyses are performed using absorption of light as the analytical tool using an instrument known as a spectrophotometer (Experiment 29 - Spectrophotometric Determination of Iron in a Vitamin Tablet). This is but one example of instrumental analysis, in which a variety of instruments may be employed, depending on the type of analysis being done.

Qualitative analysis, on the other hand is complicated by the lack of knowledge of what is actually present, thus the analyst has to look for a variety of different possibilities, and be able to conclusively rule out those species that are not present while identifying those that are present. Unlike quantitative analysis, however, qualitative analysis is not concerned with exact amounts of substances present, only whether they are there or not. When a farmer's cows are sick or dying, the farmer needs to know quickly whether a stream flowing through his field is contaminated, as well as the identity of the contaminant. He is not concerned with the exact concentration of the toxic substance. When a demented person put several contaminated containers of a popular anti-inflammatory medicine on the shelf of a pharmacy a few years ago, and caused several deaths as a result, it was important to identify *what* poison was present, not the exact amount in the bottle.

The type of procedures employed for qualitative analysis, as for quantitative analysis, depends on what the nature of the unknown materials may be. This section of the laboratory manual is concerned with analysis of ionic materials, through identification of the individual ions. Qualitative identification of ions is normally done in aqueous solution, where the ionic substances are completely dissociated and the ions are independent of one another. This independence allows one to analyze for the cations regardless of the anions present, and vice versa. The cations present in ionic materials are almost invariably metal ions (with the exception of the ammonium ion, NH_4^+, which is the only commonly encountered polyatomic cation), while the anions run the gamut from monatomic (such as the halides - Cl^-, Br^-, I^-) through the many polyatomic ones (such as sulfate – SO_4^{2-}, carbonate - CO_3^{2-} etc). Inorganic qualitative analysis is the identification of the ions present in a solid ionic material or in an aqueous solution.

Two types of responses to chemical tests give the majority of information necessary for the identification of cations. These are (1) the appearance of color and (2) solubility considerations - that is,the appearance of precipitates. For identifications of anions, color change is an aid, but formation of precipitates and evolution of gases, along with the identification of the gases formed provide the most important information. For all testing procedures, **IT IS OF PARAMOUNT IMPORTANCE TO CAREFULLY OBSERVE AND TO RECORD OBSERVATIONS.** Detailed records of the tests run and of the results of the tests must be maintained, both to provide the information necessary to identify the unknown and to prevent confusion with results. For example, if one adds silver nitrate to an unknown to check for the presence of chloride ion, then one can be assured that the unknown will test positive for silver at a later time, even if silver had not been originally present. Obviously, one must know what has been added to the solution at all times!

Analysis of a mixture consists of two general processes – separation and identification. Separation of unknown ions is necessary for the same reason that the color of a paint sample is not indicative of the pigments added to the bucket. Paint may be green because one pure green pigment was added or because both blue and yellow pigments were added. The eye perceives only the resultant, not the components. The copper (II) ion (Cu^{2+}) gives an intense royal blue color when a solution of aqueous ammonia (also called ammonium hydroxide) is added, due to the formation of the $Cu(NH_3)_4^{2+}$

complex ion, and this is a confirmatory test for the presence of copper. Iron (III) ion (Fe^{3+}), on the other hand, produces a reddish brown jello-like precipitate of $Fe(OH)_3$ in the presence of ammonia. This is a confirmatory test for the presence of iron. If *both* copper and iron are present in the same solution, the two tests interfere with one another, because one can see only the result of mixing a blue color with a red-brown precipitate, and it is impossible to interpret the resultant mixture a confirmatory for either. Clearly, one must first *separate* the copper and iron, and then *confirm* each independently.

Separation is accomplished using what are known as "group solubilities." Table 18-1 lists the common group solubilities in qualitative analysis.

TABLE 18.1

GROUP	SOLUBILITY PATTERNS
Ag^+, Pb^{2+}, Hg_2^{2+}	Chloride (Cl^-) salts are insoluble
Cu^{2+}, Hg^{2+}, Bi^{2+}, Sn^{4+}, Cd^{2+}, As^{3+}	Sulfide (S^{2-}) salts are *insoluble in strongly acidic solution*
Ni^{2+}, Fe^{2+}, Fe^{3+}, Cr^{2+} Co^{2+}, Mn^{2+}, Zn^{2+}	sulfide (S^{2-}) salts and hydroxide (OH^-) salts are *insoluble in strongly basic solution*
Ca^{2+}, Ba^{2+}	carbonate (CO_3^{2-}) salts are insoluble
NH_4^+, Na^+, K^+	All salts are soluble

The use of information included in the tables is as follows. Imagine being given a solution which might contain any combination of the cations NH_4^+, Ag^+, Pb^{2+}, Hg_2^{2+}, Cu^{2+}, Fe^{3+}, Ca^{2+}, Na^+, Cd^{2+}, Bi^{3+}, and Ni^{2+}. One must conclusively show which ions are present, and conclusively rule out the ones that are absent. This requires several separation steps, and a confirmation step for each ion. We will discuss the complete separation and identification of one group of ions (the silver group), with the understanding that similar analyses of additional groups utilize the same general principles, but involve different reagents and observations.

Since almost all chloride-containing ionic materials are fairly soluble **WITH THE EXCEPTION OF SILVER (Ag^+), LEAD (Pb^{2+}) AND MERCURY (I) (Hg_2^{2+}*) CHLORIDES**, then this group of ions (known as the silver group) can be separated from all other metal ions in solution through the addition of a source of chloride ion to the unknown solution. The source of chloride ion chosen is hydrochloric acid, HCl_{aq}, since no additional metal cations are added at the same time. If any combination of the silver group metal ions is present in the solution, then a precipitate will form which includes $AgCl$ and/or $PbCl_2$ and/or Hg_2Cl_2. Since all of these are white solids, one cannot tell by looking which one or which combination is present. However the ions in this *group* may now be separated from all other ions potentially present in the initial solution merely by filtering the solution and isolating the precipitate. The filtrate contains the remainder of ions to be identified, and would be set aside, *clearly labeled*, and saved for further analysis.

One now sets out to separate and identify the components of the silver group actually present in the white precipitate. This is done by focusing on the specific properties of the substances potentially comprising the white precipitate. Lead chloride ($PbCl_2$) is soluble in hot water, but insoluble in cold water. The other two metal chlorides of the silver group are insoluble, even in hot water. It becomes a simple matter, then to treat the white precipitate with hot water, and then filter the mixture *while it is very hot*. The lead chloride will be contained in the filtrate, and the silver and mercury chlorides will be left as white residue on the filter paper. Lead chloride is colorless in water, thus the filtrate will look like

---

* The Hg_2^{2+} ion, also known as the mercurous ion or as the mercury (I) ion, consists of two mercury atoms bonded together, and which *together* have lost two electrons. Each mercury is thus considered to be a +1 ion, and the dimeric (two part) ion has an overall charge of +2. This ion is not to be confused with the Hg^{2+} ion (also known as the mercuric ion or the mercury (II) ion), which is one atom that has lost two electrons. Mercury (I) compounds have quite different properties from mercury (II) compounds.

like water, whether or not there is any lead present. One must now do a confirmatory test on the filtrate to see if lead is, indeed present. This is done through the addition of a reagent that gives a confirmatory test for lead - namely a solution of potassium chromate (K_2CrO_4). Since the compound lead chromate ($PbCrO_4$) is an insoluble, canary yellow solid, if lead is present in the unknown solution, a canary yellow precipitate will form upon the addition of the potassium chromate solution. ***The yellow color is not enough to confirm the presence of lead, since the reagent itself is yellow.*** There must be a precipitate present!

We now turn our attention to the residue remaining on the filter paper from the hot water filtration. It contains $AgCl$ and/or Hg_2Cl_2 (and nothing else, since only these chlorides, along with the lead, were insoluble). These two may be separated by making use of their individual characteristics, as lead was due to its solubility. The individual difference in this case is the fact that silver ions form a soluble complex ion with ammonia, while mercury does not. Treating the solid, still on the filter paper, with concentrated aqueous ammonia (ammonium hydroxide) will dissolve the silver (due to the formation of $Ag(NH_3)_2^{2+}$ ions) and allow it to pass through the filter paper into a collection test tube while the mercury does not. Like lead, silver ions in solution are colorless, so the collected solution will look like water whether or not the silver is present. Thus a confirmatory reagent must be added to the test tube to determine whether silver is present or not. This reagent is nitric acid, which "destroys" the ammonia and lets the white $AgCl$ precipitate reform. Therefore, the addition of HNO_3 to the text tube will produce a cloudiness or white precipitate *if silver is present.*

The final confirmation, that of the presence or absence of mercury (I), actually takes place while one is separating the silver with the aqueous ammonia. While the Hg_2^{2+} ion does not form a complex ion with ammonia, it is unstable in the presence of ammonia, and when the ammonia is added to the residue on the filter paper, the Hg_2^{2+} ions *break down* to form Hg^{2+} ions (mercury (II) ions) and mercury atoms ($Hg_2^{2+} \rightarrow Hg^{2+} + Hg$). This is called disproportionation or auto-oxidation/reduction. The Hg^{2+} ions immediately form insoluble $Hg(OH)_2$ on the filter paper, and the Hg atoms likewise are trapped on the filter paper. The combination of these two materials produces a grey-black gooey substance on the filter paper, thus this appearance when ammonia is added is confirmatory for the presence of mercury (I) ions in the original solution.

By examining Table 46-1 above, one can easily see that the filtrate from the initial separation of the silver group (which was saved and clearly labeled) would contain any other ions present in the original mixture. It is strongly acidic (the precipitation of the silver group was accomplished by the addition of hydrochloric acid), so the addition of a source of sulfide ions to this filtrate would cause the precipitation of the second group of ions listed in the table. The solid mixture containing these ions would be analyzed by a scheme designed to separate and identify based on the properties of the individual ions, as was the silver group. The filtrate from this step would contain the remaining ions. If this filtrate were now made basic (remember, you have already added sulfide ions), the third group of ions would then precipitate, and could be separated and identified. One continues through the complete analysis in this fashion, first separating groups of ions from the original solution, then working within each group to separate and identify the individual ions present.

FLOW DIAGRAM

The instructions for accomplishing the separation and confirmation steps described earlier are summarized in a schematic diagram known as a ***flow diagram***. A flow diagram provides a roadmap for qualitative analysis. It is important to be able to interpret a flow diagram, since the entire analysis can be done using only the diagram as a guide. The format for a flow diagram is quite simple, and is illustrated in Figure 18-1. One begins with a listing of the ions to be separated, at the top of the diagram. A vertical line down from this list indicates the addition of a reagent that yields a precipitate, with the identity of the reagent and its concentration written beside the vertical line. The vertical line ends in a horizontal line, which indicates the separation of the precipitate and the filtrate - which appear at opposite ends of the horizontal line. **PRECIPITATES ARE INDICATED AS NEUTRAL COMPOUNDS, WHILE SUBSTANCES REMAINING DISSOLVED IN SOLUTION ARE INDICATED AS IONS.** For example, when the HCl solution is added in step one of the figure, $AgCl$ precipitates, and Fe^{3+} remains in solution. This format continues throughout the flow diagram. As a test for correct interpretation of the flow diagram format, follow the three ions Ag^+, Cu^{2+}, and Fe^{3+} through the flow diagram shown in Figure. Upon addition of 6 M HCl, the silver separates as $AgCl$ precipitate, while Cu^{2+} and Fe^{3+} remain in solution. Addition of the compound thioacetamide, which produces sulfide ions (S^{2-}) in solution, results in the precipitation of copper sulfide (CuS), while Fe^{3+} remain in solution. Addition of the compound thioacetamide, which produces sulfide ions (S^{2-}) in solution results in the precipitation of copper sulfide (CuS), while Fe^{3+} remains in solution. When the solution is made basic by the addition of NH_4OH, the $Fe(OH)_3$ precipitates out of solution. Figure 18-1 is a general flow diagram which shows the stepwise precipitation of the various groups of ions. Each group that separates is analyzed through a more detailed flow diagram, and the individual group diagrams are shown in Figures 18-2 and 18-3.

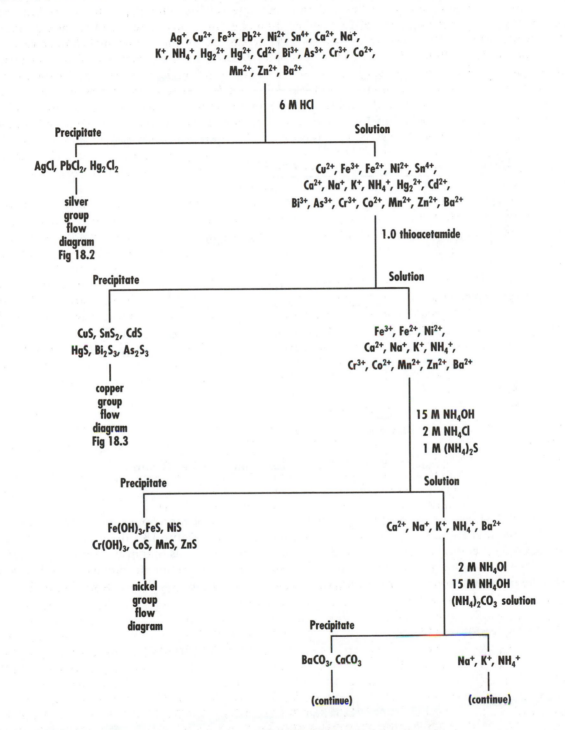

Figure 18-1. General flow diagram for analysis of all groups.

Figure 18-2 shows a more detailed flow diagram depicting the analysis of the silver group, after precipitation with the HC1. A quick summary is that the solid is treated with hot water to dissolve the $PbCl_2$, then the precipitate is treated with aqueous ammonia (NH_4OH) to dissolve the silver. Follow this flow diagram through with the extended discussion on analysis of the silver group previously.

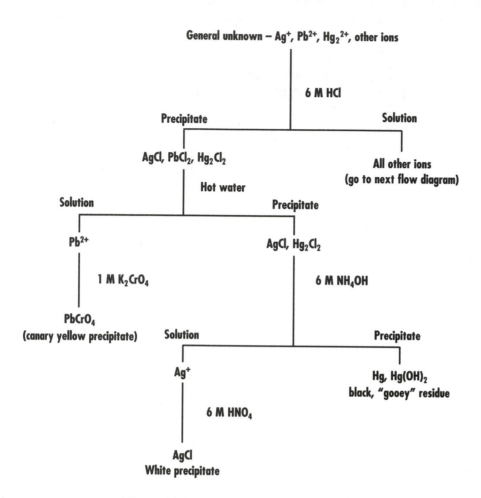

Figure 18-2. Flow Diagram for Group I (Silver Group).

Figure 18-3 shows a flow diagram for analysis of a portion of the copper group (acid insoluble sulfides). Note that Figures 18-2 and 18-3 are actually portions of Figure 18-1 the master flow diagram which describes a complete analysis procedure. One can embark on as optimistic and complicated an analytical scheme as time and interest permit. You instructor will determine the range of ions that would possible be present in your unknown, and which parts of the various flow diagrams will be used.

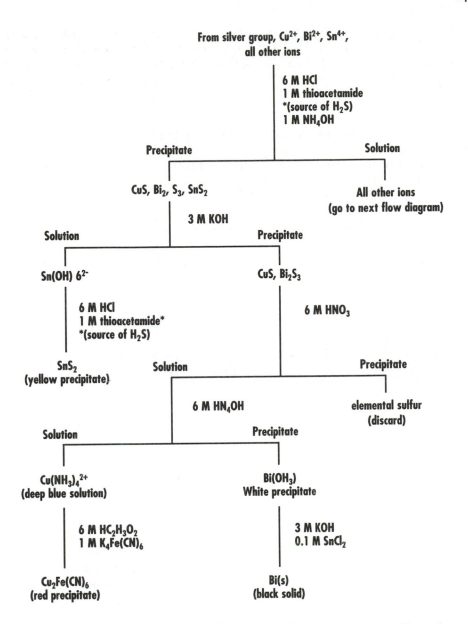

From silver group, Cu^{2+}, Bi^{2+}, Sn^{4+},
all other ions

6 M HCl
1 M thioacetamide
***(source of H₂S)**
1 M NH₄OH

Precipitate   **Solution**

CuS, Bi_2, S_3, SnS_2 **All other ions**
                                                  (go to next flow diagram)

3 M KOH

Solution   **Precipitate**

$Sn(OH)$ 6^{2-} CuS, Bi_2S_3

6 M HCl   **6 M HNO₃**
1 M thioacetamide*
***(source of H₂S)**

SnS_2 **Solution**   **Precipitate**
(yellow precipitate)

                     6 M HN₄OH   **elemental sulfur**
                                                           (discard)

Solution   **Precipitate**

$Cu(NH_3)_4{}^{2+}$ $Bi(OH)_3$
(deep blue solution)   **White precipitate**

6 M HC₂H₃O₂   **3 M KOH**
1 M K₄Fe(CN)₆   **0.1 M SnCl₂**

$Cu_2Fe(CN)_6$ $Bi(s)$
(red precipitate)   **(black solid)**

Figure 18-3. Flow Diagram for Group II (Copper Group).

An important technique that should be employed during a qualitative analysis scheme is to carry out each test *using solutions known to contain the ions in question* before beginning to analyze an unknown solution. Analysis of a known solution allows the analyst to have actually seen the positive result, and to have recorded his or her observation of what a positive test looks like. It is one thing to *read* that a "gooey black residue" forms, but it is another thing to have seen it and recorded personal observations of what it actually looks like *to you.* Remember, *you* are the only one who will decide whether a given test does or does not indicate the presence of a particular ion in your unknown!

PROCEDURE

As you proceed through the experimental procedure, there are several principles and general procedures that you must remember at all times. These apply to all parts of the analysis.

1. This is semi-micro analysis, which means that very small quantities of solutions are being used. Nevertheless, many metal ions are quite toxic, thus **NO SOLUTIONS SHOULD BE DISCARDED DOWN THE DRAIN WITHOUT THE PERMISSION OF YOUR INSTRUCTOR.** Waste solution containers will be maintained for disposal of all solutions containing

metal ions. The metals in these general waste solutions can be precipitated together, in one easy step, for safe disposal at the end of the laboratory exercise.

2. *The order of analysis is very critical.* Groups are precipitated in order of increasing solubility. This means that if group I ions are not precipitated first, then they will precipitate along with the group II ions and make analysis of the group II ions more difficult. Stepwise precipitation of the groups is much like separating a pile of various sized rocks into sized groups. One begins with a screen with small holes, through which only small rocks can pass. Successive use of larger screens allows collection of larger sizes of rocks. If one began with a screen with large holes, no separation would be accomplished, as the small and large rocks would all go through. So one precipitates groups in *increasing* order of solubility.

3. The most frequent reason for incorrect analysis is contamination of the sample. Use only distilled or deionized water when making solution or rinsing glassware, and *double check the label on each reagent before you use it.* **ALL GLASSWARE SHOULD BE CLEANED AND RINSED PRIOR TO EACH USE.** This last statement especially applies to stirring rods used to mix solutions.

4. Another common reasons for error is getting mixed up with the procedures or observations. **RECORD YOUR OBSERVATIONS EACH TIME YOU RUN A TEST, AND WRITE AS MUCH ABOUT THE TEST AS IT WILL TAKE FOR YOU TO REMEMBER THE APPEARANCE OF THE RESULTS AT A LATER TIME.** Complete record keeping is as important as running the tests properly.

5. Very frequently you will be required to separate precipitates from the liquids from which they formed. After the precipitates have been isolated, they must be thoroughly washed each time, in order that they are not contaminated by dissolved ions remaining in the solution which wets the precipitate after filtration.

6. Sometimes the separation of the precipitate and filtrate will be done using a centrifuge. This device spins the test tube very rapidly and forces the heavier precipitate particles to the bottom of the test tube. *When the centrifuge is used, it must be balanced at all times with a test tube containing an equal volume of water directly across from the sample test tube!* When washing a precipitate that has been centrifuged, it is important to stir the precipitate in the wash solution and completely resuspend it in the solution before centrifuging and discarding the wash solution.

7. You will use many solutions of acids and bases during the qualitative analysis scheme. These are all potentially quite damaging to eye tissue, particularly when hot. Solutions can spatter out of test tubes while being heated, thus, **SAFETY GOGGLES MUST BE WORN AT ALL TIMES IN THE LAB.**

8. All analyses should be done twice - once with a solution that is known to contain the ions being tested for, and a second time on a solution provided by your instructor which may contain any or all of the ions. Your instructor may provide the known solution, or you can prepare it yourself from solutions containing the ions. The instructions below apply to both the known solution and the unknown solution.

Separation and Analysis of Group I (the Silver Group)

1. Place about 10 drops of the solution to be analyzed into a small test tube and add 6 drops of 6 M HCl. The appearance of a white precipitate indicates the presence of one or more ions from the silver group. If no precipitate is seen, then no ions of the silver group are present and you may proceed directly to analysis of Group II (the copper group), using the solution to which you added the 6 M HCl. If a white precipitate *does* form, proceed with the steps below.

Chemistry: $Ag^+, Pb^{2+}, Hg_2^{2+} + xs\ Cl^- \rightarrow AgCl, PbCl_2, Hg_2Cl_2$ *

> * The abbreviation xs stands for "excess."
>
> * In all equations to follow, a species written as an ion (with a charge) is dissolved in the solution, while a species written as a neutral compound is present as a precipitate or a gas.

2. Centrifuge the mixture for 2 minutes, and carefully decant (pour off) the liquid (hereafter called the supernatant or supernatant liquid) into a clean test tube. If the known or unknown contains only the silver group ions, this solution may be discarded. If the solution may contain ions from other groups, this solution should be clearly labeled and set aside for further analysis following completion of the silver group analysis.

3. Wash the precipitate twice with 2 or 3 mL of deionized or distilled water *to which 3 drops of 6 M HCl have been added.* For each washing, stir the solution well to resuspend the precipitate before centrifugation. Pour off the last wash solution, leaving only the wet precipitate in the test tube.

4. Add 15 drops of distilled water to the precipitate, stir the mixture thoroughly, and place the test tube in a boiling water bath for 5 minutes, stirring occasionally. Carefully remove the test tube from the boiling water bath and *quickly* filter the hot solution, catching the filtrate in a clean test tube. The presence of white residue on the filter paper is indicative of the presence of mercury (I), silver, or both.

Chemistry: $PbCl_2 + hot\ water \rightarrow Pb^{2+} + 2\ Cl^-$

5. To the clear filtrate from step 4, add 4 drops of 1 M potassium chromate (K_2CrO_4) solution. The appearance of a canary yellow precipitate confirms the presence of lead.

Chemistry: $Pb^{2+} + CrO_4^{2-} \rightarrow PbCrO_4$

6. Wash the residue on the filter paper from step 4 with 6 drops of 6 M NH_4OH, making sure to contact all of the residue with the base, catching the solution in a clean test tube as it goes through the filter paper. The appearance of a grey-black residue on the filter paper indicates the presence of mercury (I).

Chemistry: $Hg_2Cl_2 + xs\ OH^- \rightarrow Hg\ (OH)_2 + Hg$

$AgCl + 2\ NH_3 \rightarrow Ag\ (NH_3)_2^{2+} + Cl^-$

7. Add 6 M HNO_3 dropwise to the solution from step 6 until the solution tests acidic to litmus paper (it will take 8 or more drops). The appearance of a white cloudy precipitate indicates the presence of silver.

Chemistry: $Ag\ (NH_3)_2^{2+} + Cl^- + 2\ H^+ \rightarrow AgCl + 2\ NH_4^+$

Analysis of Group II (the Copper Group)

Only a limited part of the analysis of the copper group is used in this scheme. The copper group actually contains about 10 different metal ions, and only 3 are encountered in this analysis scheme. This abbreviated analysis will demonstrate the principles of separation and identification without occupying undue laboratory time that will be used for other types of investigations.

1. Begin this analysis with the solution saved after precipitation of group I (the silver group, or with 10 drops of a solution known to contain group II ions. In the latter case, you must add 6 drops of 6 M HCl. You have already added this acid to the solution from group I.

2. Add 10 drops of 1 M thioacetamide (CH_3CSNH_2) solution to the solution from step 1, and allow the test tube to sit in a boiling water bath for 10 minutes. Remove the solution from the water bath, cool it under running water, and add 1 drop of 1 M NH_4OH. This promotes the precipitation of the SnS_2. The appearance of precipitate at this point indicates the presence of one or more ions of the copper group. If no precipitate is present, then no ions of this group are present.

Chemistry: **The thioacetamide breaks down in the aqueous solution to produce hydrogen sulfide (equation 1). The hydrogen sulfide ionizes to produce first hydrogen sulfide ions (equation 2), then sulfide ions (equation 3). The metal ions precipitate in the presence of the sulfide ions (equation 4).**

$$CH_3CSNH_2 + H_2O \rightarrow CH_3CONH_2 + H_2S \tag{1}$$

$$H_2S \rightarrow H^+ + HS^- \tag{2}$$

$$HS^- \rightarrow H^+ + S^{2-} \tag{3}$$

$$Cu^{2+}, Sn^{4+}, Bi^{3+} + xs\ S^{2-} \rightarrow CuS, SnS_2, Bi_2S_3 \tag{4}$$

Hydrogen sulfide has a particularly obnoxious odor (like rotten eggs) and it is also fairly toxic. For these reasons it is generated in the reaction mixture as shown above, rather than being used as a gas. This keeps the concentration very low, and the gas dissolved in solution where it is relatively harmless. Generating a reactant in solution like this is a common synthesis technique, and is known as generating the reactant *in situ* (Latin for "in the position").

3. Centrifuge the test tube, and without pouring off the supernatant liquid, add 5 additional drops of thioacetamide and again place the test tube int he hot water bath for 3 more minutes. Cool as before and again add 1 drop of the aqueous ammonia. Stir the solution and centrifuge again. This time, the supernatant liquid may be decanted. If you were doing a general unknown, which might contain ions from groups III and beyond, you would label this liquid and save it for future analysis. Since no further groups will be investigated, you may discard this solution down the drain.

4. Wash the precipitate by adding 15 drops of 0.1 M NH_4Cl solution and stirring the mixture. Centrifuge the mixture and discard the supernatant. The precipitate contains the group II metals to be identified.

5. To the precipitate from step 4, add 20 drops of 3 M KOH and 1 drop of 1 M thioacetamide. Stir the mixture, heat the test tube in a hot water bath for 2 minutes and centrifuge. Decant the supernatant liquid into a clean test tube. If the original solution contained tin, it will be present in this solution. The CuS and Bi_2S_3 are insoluble in this solution. If a precipitate is still present at this point, the presence of copper and/or bismuth is indicated. Save this precipitate for further testing.

Chemistry: **The tin is made soluble through the formation of a complex ion. The added thioacetamide keeps the other two sulfides from dissolving.**

$$SnS_2 + xs\ OH^- \rightarrow Sn\ (OH)_6{}^{2-} + 2\ S^{2-}$$

6. To the supernatant liquid from step 5, add 6 M HCl dropwise until the solution tests acidic to litmus paper (at least 12 drops). Then add 5 drops of 1 M thioacetamide solution. Heat the test tube in a hot water bath for 3 minutes. A yellow/brown precipitate indicates the presence of tin (Sn).

Chemistry:

$$Sn(OH)_6^{2-} + xs\ H^+ \rightarrow Sn^{4+} + 6\ H_2O$$
$$Sn^{4+} + 2\ S^{2-} \rightarrow SnS_2$$

7. Wash the precipitate from step 5 two times, each time adding 10 drops of distilled water, stirring thoroughly, and centrifuging. Both times the washing liquid may be discarded. Add 10 drops of 6 M HNO_3 to the precipitate and heat the test tube in the boiling water bath for 5 minutes. The precipitate should dissolve, with the possible formation of a second, yellow residue during this step. This residue is elemental sulfur and should be discarded. The solution contains the copper and/or bismuth ions.

Chemistry:

$$2\ H^+ + CuS + 2\ HNO_3 \rightarrow Cu^{2+} + 2\ NO_2 + S + 2\ H_2O$$
$$3\ H^+ + Bi_2S_3 + 3\ HNO_3 \rightarrow 2\ Bi^{3+}\ 3\ NO_2 + 3\ S + 3\ H_2O$$

8. To the solution from step 7, add 6 M NH_4OH dropwise until the solution tests alkaline to litmus, then add an additional 6 drops. Thoroughly mix the contents and allow them to cool. Centrifuge the test tube. A white precipitate indicates the possible presence of bismuth, while a deep blue supernatant solution indicates the presence of copper.

Chemistry:

$$Cu^{2+} + xs\ NH_3 \rightarrow Cu(NH_3)_4^{2+}$$
$$Bi^{3+} + xs\ OH^- \rightarrow Bi(OH)_3$$

9. Decant the supernatant solution from step 8 into a clean test tube. If it is colorless, no additional tests are warranted, as copper is definitely absent. If the solution is blue, transfer 3 drops of the solution into a clean test tube and add 10 drops of 6 M $HC_2H_3O_2$ and 5 drops of 1 M $K_4Fe(CN)_6$. Centrifuge the test tube and decant the liquid. A red precipitate confirms the presence of copper.

Chemistry:

$$2\ Cu(NH_3)_4^{2+} + Fe(CN)_6^{4-} \rightarrow Cu_2Fe(CN)_6 + 4\ NH_3$$

10. If there is a precipitate from step 8, the presence of bismuth is strongly indicated. Wash the precipitate twice with 10 drops of distilled water to which 4 drops of NH_4OH have been added. Stir the mixture and centrifuge each time. Discard the final wash solution, then add 10 drops of 3 M KOH and 5 drops of 0.1 M $SnCl_2$ solution to the precipitate. Stir the solution well and allow it to sit for 3 minutes. The production of a black solid (elemental bismuth) confirms the presence of bismuth.

Chemistry:

$$2\ Bi(OH)_3 + 3\ SnCl_2 \rightarrow 2\ Bi + 3\ Sn^{4+} + 6\ Cl^- + 6\ OH^-$$

11. Clean and rinse all glassware thoroughly before putting it away. Make sure to clean up the work station well, as the acids and bases used in this experiment are very corrosive.

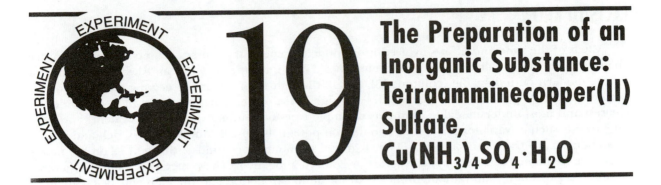

19 The Preparation of an Inorganic Substance: Tetraamminecopper(II) Sulfate, $Cu(NH_3)_4SO_4 \cdot H_2O$

OBJECTIVE

To learn something about coordination compounds by preparing one in the laboratory, watching it undergo chemical change, and partially characterizing that change.

APPARATUS AND CHEMICALS

250-mL beaker
150-mL beaker
100-mL graduated cylinder
125-mL conical flask
glass stirring rod
vacuum filter flask

Büchner funnel
filter paper
anhydrous copper sulfate
copper sulfate pentahydrate
concentrated ammonia
ethanol

SAFETY CONSIDERATIONS

 Copper sulfate is a water-soluble poison. Even though the copper ion is required in some plant and animal metabolic processes, large amounts of the ion in living tissue are dangerous. Dispose of waste material and clean up as directed by your instructor. Be sure to wash your hands at the end of the laboratory.

 Concentrated ammonia is corrosive as a liquid, and the ammonia gas that is given off from the saturated solution is irritating to the eyes and nose. Do not open a container of this chemical without adequate ventilation. Also, you should protect good clothing from droplets or gas from concentrated ammonia.

 The ethanol used in this experiment is denatured in that a poison has been added so that it will not come under the legal controls associated with beverage alcohol.

 Be sure to wear eye protection when working in the chemistry laboratory.

FACTS TO KNOW

Much of the time and effort involved in introductory chemistry courses centers on the chemistry (behavior) of the so-called "main group elements." These are the first two and last six columns of on the modern periodic table (columns 1,2, and 13 through 18 on the most recent generations of the periodic chart, or columns IA, IIA, and IIIA through VIIIA on the older versions). They provide the clearest and most understandable examples of simple types of bonding (ionic and covalent) and periodic trends such as variations in size and ionization potentials. Discussion of the very rich and important chemistry of the *transition metals* (The "B-group" elements or columns 3 through 12 on the newer charts) is frequently touched on very lightly or ignored altogether. These metals—Scandium (atomic number 21) through Zinc (atomic number 30) in the fourth period and the elements below them in periods 5 and 6—frequently form compounds in which the metal ions form bonds with certain neutral molecules or anions to produce large *complex ions* or neutral *metal complexes* that have some properties quite unlike those of the more "traditional" compounds of these elements. Frequently these compounds are very highly colored, with hues ranging throughout the range of colors in the visible spectrum. Iron (Fe), for example, frequently forms yellow or green complexes, with some being red or blue, while the most common colors for copper (Cu) complexes are blues and purples. Some of these complexes are very good oxidizing or reducing agents (meaning they can easily accept or donate electrons to other species).

The element copper serves to illustrate some features of these complexes. A commonly encountered compound of copper is copper sulfate ($CuSO_4$). One can obtain either anhydrous copper sulfate (anhydrous means no water is present), or copper sulfate pentahydrate. The term "hydrate" means that water molecules are a part of the crystalline structure of the material (see Experiment 12 - Determination of the Formula of a Hydrate). The difference between these two substances (both are copper sulfate) is remarkable. The anhydrous copper sulfate is a soft white powder that gets warm and turns blue when water is added. The copper sulfate pentahydrate ($CuSO_45H_2O$) is a bright blue crystalline substance that is commonly called "bluestone" because it looks like clear blue gravel. It does not get hot when water is added to it. (The anhydrous copper sulfate is sometimes used as a test for the presence of water, due to the production of the blue color with water.)

Anhydrous copper sulfate is a classic example of an ionic compound, exactly like sodium sulfate or calcium sulfate. These consist of a lattice of positive ions (Cu^{2+}, Na^+ or Ca_2^+) and negative ions (SO_4^{2-}) in the familiar repeating pattern of cations surrounded by anions and vice versa. In the case of copper sulfate pentahydrate, the water molecules are a part of the crystal lattice as well, but the manner in which they interact with the copper is what makes the compound somewhat unique. The water molecules form bonds with the copper ion—this is called coordination. A larger ion is thus created in which the 2+ charge of the copper ion is "spread out" over a much larger *complex ion* $[Cu(H_2O)_4]_2^+$ which occupies a much larger position in the ionic lattice with the sulfate ions (Figure 19-1). The complex ion has a 2+ charge because the water molecules are neutral, while the copper ion retains its charge. The blue color is due to the interaction of the water molecules with the copper.

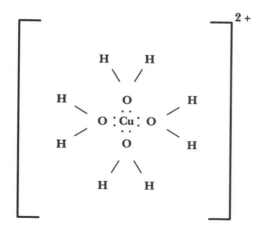

Figure 19-1

Coordination complexes form because transition metal ions have empty orbitals (3d in the case of copper) of a low enough energy that they can form bonds using these orbitals by *accepting electron pairs donated by other molecules or ions*. It is important to understand that the electrons are NOT simply transferred to the metal ion—this would be oxidation/reduction—but are shared by the metal and donor molecule (or ion) *to form a bond*. There is a large class of these coordination complexes (also called coordination compounds) in which the metal ion is bonded to several other molecules or ions by means of these coordinate bonds (also called "dative bonds"). A coordinate bond is formed when both of the shared electrons forming the bond come from the same atom. Consider the title compound for this experiment, tetramminecopper(II) sulfate. Copper is bonded to four ammonia molecules by four electron pair bonds made by the sharing of an electron *pair* from each ammonia with the copper ion (Figure 19-2).

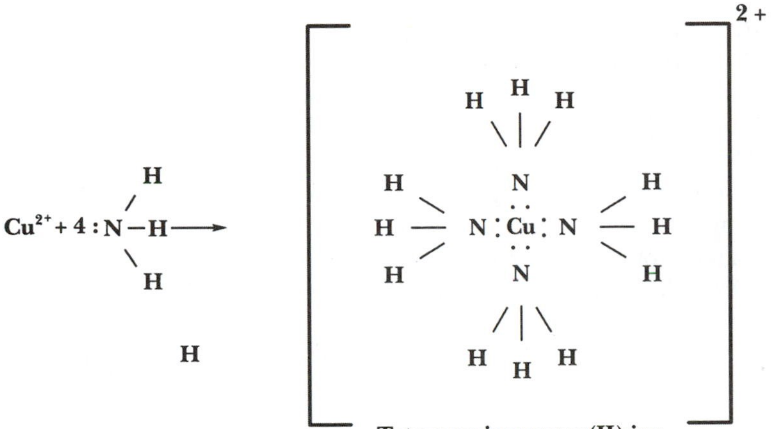

Tetraamminecopper(II) ion

Figure 19-2

Coordination compounds may form with metals other than transition metals (see the example with magnesium below) but they are most commonly encountered with transition metals. While coordination compounds may seem to be rather esoteric curiosities which are not discussed much in class, they actually play many vital roles in your everyday lives. Chlorophyll, the green pigment that enables plants to produce oxygen for you to breathe is a coordination compound of magnesium. Hemoglobin, the component of your blood that carries the oxygen to your cells is a coordination compound of iron (Figure 19-3).

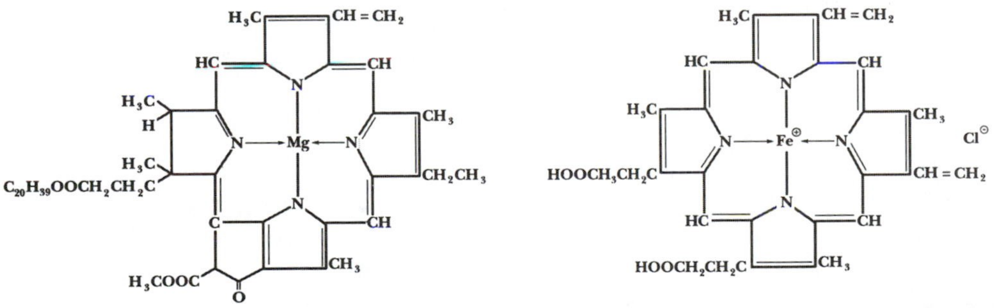

chlorophyll

hemin

Figure 19-3

As mentioned above, one obvious physical characteristic of coordination compounds is that of color. The green of chlorophyll, the red of hemoglobin, and the blue of bluestone are examples. As you proceed with this experiment, pay particular attention to the color changes observed, beginning with the anhydrous copper sulfate and ending with the isolation of the powdery crystals of tetramminecopper (II) sulfate as your final product. Remember not to discard any of your copper-containing materials down the drain, but dispose of them as instructed by your laboratory supervisor.

PROCEDURE

Numbers in parentheses refer to entry numbers on data sheet.

I. PREPARATION OF COMPLEX dissolve 2.0 g of anhydrous copper sulfate ($CuSO_4$) in 15 mL of distilled water by adding the powder slowly with stirring. Use a glass stirring rod. Note the color of the solution (1). This color is due to Cu^{2+} ions being coordinated by water molecules. The instructor will have on display some solid hydrated copper sulfate, $CuSO_4 \cdot 5H_2O$. Compare the color of your solution with the solid $CuSO_4 \cdot 5H_2O$.

To this solution slowly add 7 mL of concentrated ammonia. Use a hood at this point and be sure to stir the liquid.

The first addition of NH_3 causes the light blue copper hydroxide, $Cu(OH)_2$, to be formed. This dissolves to form tetraamminecopper(II) ions, $Cu(NH_3)_4^{2+}$, upon addition of excess NH_3.

Now slowly add 25 mL of ethanol to the $Cu(NH_3)_4^{2+}$ solution with stirring. The precipitate formed is $Cu(NH_3)_4SO_4 \cdot H_3O$.

Allow the precipitate to stand for 5 minutes and then filter. Set up a vacuum filtration apparatus (Fig. 19-4) with a Buchner funnel and place a filter paper in the funnel. Moisten the filter paper with a small amount of water from your wash bottle and turn on the vacuum to seat the filter paper. To ensure a good yield, be certain to scrape all the product from the walls of the beaker with a rubber policeman on a glass rod as you filter the solution.

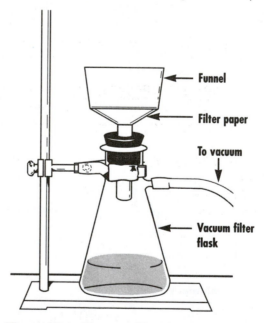

Figure 19-4. Vacuum filtration apparatus.

Wash the precipitate with two 5 mL portions of ethanol. Before adding each wash portion, turn off the vacuum. Then pour in 5 mL of ethanol, stir carefully to ensure contact of the ethanol with the solid, then turn on the vacuum to remove the ethanol. Repeat the same procedure with the second wash portion. After the ethanol from the second wash has been pulled through, leave the vacuum on until the solid appears to be dry (5 or 10 minutes).

Weigh the dry product on a preweighed watch glass (4 and 5). Record the weight of your product (6).

II. SOLUBILITY TEST   Weigh out about 0.10 g of the dried product that you have just prepared and place it in about equal amounts in two dry test tubes. Add water to one test tube until it is half filled and add an equal amount of alcohol to the other test tube. Shake both to ensure good mixing. Describe the results (7).

III. CALCULATIONS AND REPORT The reaction for the preparation of $Cu(NH_3)_4SO_4 \cdot H_2O$ is

$$CuSO_4 + 4\,NH_3 \xrightarrow[H_2O]{in} Cu(NH_3)_4SO_4 \cdot H_2O$$

From the balanced chemical equation we see that one mole of $CuSO_4$ yields one mole of $Cu(NH_3)_4SO_4 \cdot H_2O$. Based on the weight of your product (6), calculate the percent yield (8).

$$\text{Wt } Cu(NH_3)_4SO_4 \text{ (theory)} = \text{wt } CuSO_4 \text{ taken} \times \left[\frac{\text{formula wt } Cu(NH_3)_4SO_4 \cdot H_2O}{\text{formula wt } CuSO_4} \right]$$

$$\%\text{ yield} = \frac{\text{wt } Cu(NH_3)_4SO_4 \cdot H_2O \text{ (recovered)}}{\text{wt } Cu(NH_3)_4SO_4 \cdot H_2O \text{ (theory)}} \times 100\%$$

EXAMPLE CALCULATION

8.00 g $CuSO_4$ taken
12.0 g $Cu(NH_3)_4SO_4 \cdot H_2O$ recovered
molecular weight of $CuSO_4$ = 159.6
molecular weight of $Cu(NH_3)_4SO_4 \cdot H_2O$ = 245.7

wt $Cu(NH_3)_4SO_4$ (theory) =

$$8.00 \text{ g } CuSO_4 \times \frac{245.7 \text{ g } Cu(NH_3)_4SO_4 \cdot H_2O}{159.6 \text{ g } CuSO_4} = 12.3 \text{ g}$$

$$\%\text{ yield} = \frac{12.0 \text{ g } Cu(NH_3)_4SO_4 \cdot H_2O \text{ (recovered)}}{12.3 \text{ g } Cu(NH_3)_4SO_4 \cdot H_2O \text{ (theory)}} \times 100\% = 97.6\%$$

Turn in your product to the instructor with the following information on the label:
    Name of student
    Name of product
    Actual yield in grams
    Percent yield

PRE-LAB QUESTIONS

1. Where does the term "coordination" in the phrase "coordination complex" come from (what does it mean)?

    _____

    _____

    _____

2. Give 2 examples of coordination complexes that help mammals to live.

    _____

    _____

    _____

3. What is a common *physical* characteristic of coordination complexes that can be observed?

    _____

    _____

    _____

4. What very strong smelling reactant is added to the solution of copper sulfate in water, and what precautions are to be taken while adding this reagent?

    _____

    _____

    _____

Experiment 19   **Name** _____

Data Sheet   **Date** _____

THE PREPARATION OF AN INORGANIC SUBSTANCE

(1) Weight of empty beaker   _____ g

(2) Weight of beaker plus anhydrous $CuSO_4$ _____ g

(3) Weight of anhydrous $CuSO_4$ (subtract 1 from 2)   _____ g

(4) Weight of empty watch glass   _____ g

(5) Weight of watch glass plus $Cu(NH_3)_4SO_4 \cdot H_2O$ _____ g

(6) Weight of $Cu(NH_3)_4SO_4 \cdot H_2O$ (subtract 4 from 5) _____ g

(7) Comparison of the solubility of $Cu(NH_3)_4SO_4 \cdot H_2O$ in water and ethanol

_____

_____

_____

(8) Percent yield   _____ %

Do calculations in an organized fashion below:

POST-LAB QUESTIONS

1. Describe the color changes that the system underwent, beginning with the anhydrous copper sulfate and ending with the final product.

    _____

    _____

    _____

2. What happened to the copper ions in the solution when the ammonia was added to the solution?

    _____

    _____

    _____

3. What was the purpose of adding the ethanol (ethyl alcohol) to the solution of tetramminecopper (II) sulfate?

    _____

    _____

    _____

4. Why did the change noted in question occur? How does that affect the decision as to how to wash the final product in the Buchner funnel?

    _____

    _____

    _____

EXPERIMENT 20

Chemical Reactions of Copper

OBJECTIVE

To observe a sequence of reactions of copper that start with elemental copper and end with elemental copper.

APPARATUS AND CHEMICALS

15 × 125 test tube
10 mL graduated cylinder
4 mm stirring rod
test tube rack
250 mL beaker
granular zinc
acetone
deionized water
pH Hydrion paper
scissors
ice
insulated gloves or finger covers
two dram vial

24 gauge copper wire (8 cm per student)
concentrated nitric acid
8.0 M NaOH
3.0 M H$_2$SO$_4$
6 M HCl
Pasteur pipettes with bulb
centrifuge
analytical balance (±0.001 g or better)
laboratory balance (±0.1 g)
Büchner funnel
vacuum filter flask
filter paper for Büchner funnel
drying oven

SAFETY CONSIDERATIONS

 The acid and base solutions used in this experiment can cause permanent eye damage if splashed into the eyes. Wear your safety goggles at all times. If some reagent does get into your eyes, immediately call for help from your instructor and begin flushing your eyes with cold water. If acid or base is spilled on the skin, rinse immediately with water.

 The reaction of copper wire with concentrated nitric acid must be carried out in an adequately ventilated hood because of the release of toxic vapors of nitrogen dioxide.

 The Bunsen burner is used in this experiment. Care should be exercised in lighting and using the burner. Do not forget to turn the gas off after use.

FACTS TO KNOW

Elemental copper reacts with concentrated nitric acid (HNO_3) to form copper(II) nitrate, $Cu(NO)_3)_2$. The addition of sodium hydroxide to a copper(II) nitrate solution precipitates copper(II) hydroxide, $Cu(OH)_2$. When copper(II) hydroxide is heated, it forms copper(II) oxide, CuO. Addition of sulfuric acid to copper(II) oxide produces copper sulfate, $CuSO_4$. Reduction of copper(II) sulfate with zinc gives elemental copper. This series of reactions involves four important types of chemical reactions: i) reduction/oxidation(redox); ii) acid/base; iii) precipitation, and iv) decomposition. The driving forces behind these reactions are briefly described as follows:

Reduction/Oxidation or Redox:

Electron transfer occurs from species that are electron rich (reducing agents) to species that are electron poor (oxidizing agents). In this experiment, elemental copper (reducer) is oxidized by nitric acid to form Cu^{2+}(aq) ions. Interestingly, Cu^{2+}(aq) ions can serve as oxidizers as revealed by their reaction with Zn metal, a strong reducer.

Acid/Base:

Acids are H^+ donors and bases are H^+ acceptors. Acid/base reactions are hydrogen (proton) transfer reactions. In this experiment, solid CuO is the base (due to oxide ions in $Cu^{2+}O^{2-}$) and sulfuric acid is the acid (due to H^+ in H_2SO_4). The oxide ions gain H^+ to produce water, H_2O.

Precipitation:

Charged species are unstable in nature in the sense that they seek opposite charges. This may be termed charge neutralization. In chemistry, charged ions can interact with oppositely charged ions and form an ionic solid. The cations and anions are held together by the resulting lattice energy. Charged ions can also be surrounded by polar molecules such as water molecules and thus have their charges counteracted. This involves solvation energy. Precipitates form when lattice energies of solids are larger than solvation energies of ions.

Decomposition:

Many substances can be heated to bring about chemical changes. Often gaseous products are driven off. In many instances a redox reaction actually takes place. One type of decomposition involves dehydration of metal hydroxides. Metal oxides are base anhydrides. This means that metal oxide is a dehydrated base and, if hydrated, will produce the metal hydroxide. Thus, CuO is dehydrated $Cu(OH)_2$.

PROCEDURE

Numbers in parentheses refer to entry numbers on the data sheet. To save time start the hot water bath for Step IV while you are waiting for all the copper wire to react in Step II.

I. Weigh a weighing boat on an analytical balance and record its weight (2). Cut an 8 cm length of #24 gauge copper wire into small pieces about 0.5 cm long with a scissors. Place them in the weighing boat and weigh on an analytical balance. Record the weight (1).

II. Transfer all the pieces of copper wire to a 15 × 125 test tube. *Take the test tube to the hood and carefully add 25 drops of concentrated nitric acid.* The reaction of copper with concentrated nitric acid is very vigorous at first and brown fumes of nitrogen dioxide gas, NO_2, are released. Avoid breathing these fumes. After about five minutes, most of the copper wire has reacted. However, it takes close to twenty minutes for the last bits of copper wire to react completely. Often the bubbles and the last bit of wire are so small that they are easy to overlook. If some of the wire still remains after ten minutes, add five more drops of concentrated nitric acid. (It is important to avoid adding excess nitric acid since the excess acid will require neutralization in a later step.) After you are sure production of nitrogen dioxide gas has ceased and all of the copper wire has dissolved, return to your desk. The test tube contains your first product—copper(II) nitrate. For purposes of equation writing, the reactants are Cu, HNO_3(aq), and the products are $Cu(NO_3)_2$(aq), NO_2(g) and H_2O(l). Write a balanced chemical equation for this reaction (8).

III. Add 1 mL of deionixed water to the test tube and mix. Then add 3 mL 8.0 *M* NaOH and *stir carefully* with a stirring rod. Test the solution with pH Hydrion paper to make sure the solution is basic (pH 10). (If it isn't 10 or higher, continue to add drops of 8.0 M NaOH until the pH is 10.) Then add 1 mL of deionized water, using it to rinse any solid off the stirring rod. The precipitate is copper(II) hydroxide, $Cu(OH)_2$. Write the balanced chemical equation for this reaction (9).

IV. Put about 100 mL of tap water in a 250 mL beaker and heat with a Bunsen burner until the water is nearly boiling. Place the test tube in the hot water bath. Stir the solution in the test tube occasionally to keep the solid mixed. Make sure the water level of the hot water bath is above the solution level in the test tube. Watch for the blue copper(II) hydroxide to decompose to the black powder of copper oxide. Do not remove the test tube from the hot water bath until you are satisfied all of the copper(II) hydroxide has been converted to copper(II) oxide (about 5 to 10 minutes). Write the balanced chemical equation for this reaction (10).

V. Turn off the Bunsen burner, and use an insulated glove or finger covers to remove the test tube from the hot water bath. Place the test tube in a test tube rack, and allow the test tube to cool. While the test tube is cooling, fill another 15 × 125 test tube about half-full with water. This test tube will serve as a balance when you centrifuge your test tube that contains the copper(II) oxide.

VI. Take the two test tubes to the centrifuge and place them in slots opposite one another. Close the cover and turn on the centrifuge. As soon as the centrifuge has reached full speed, turn it off and wait for the rotating test tubes to stop. Take the test tubes back to your desk and carefully remove the liquid above the copper(II) oxide (called the supernatant liquid) with a Pasteur pipette. Then wash the product three times with 6–7 mL of deionized water, centrifuging after each wash and removing the supernatant liquid each time with a Pasteur pipette. *Carefully stir* the solution and solid during each wash, and use deionized water to wash any CuO off the stirring rod before you centrifuge the test tube.

VII. Add 3 mL of 3.0 *M* H_2SO_4 to the test tube and stir until no solid remains. The solution should be a pale-blue solution with no solid. (Add another 1 mL of 3.0 M H_2SO_4 if the solid isn't all dissolved after 5–10 minutes). This is a solution of copper(II) sulfate, $CuSO_4$. Write the balanced chemical equation for this reaction (11).

VIII. Weigh out 0.4 g of granular zinc on a laboratory balance. Add about half of this to the test tube and stir for 5 min. If the solution is still blue after stirring, add a few more granulates of zinc and stir again for about 5 minutes. Proceed in this fashion until the blue color of copper(II) sulfate has disappeared, leaving a solution of colorless zinc(II) sulfate with a reddish-brown deposit of elemental copper. Try to avoid adding too much zinc. If any unreacted granular zinc remains, take the test tube to the hood and add a few mL of 6M HCl to react with the excess zinc. The reaction between zinc metal and H^+(aq) will produce H_2 gas bubbles. The hydrogen will also reduce Cu^{2+}(aq). You *should not* proceed to the next step as long as there is H_2 evolution. Write the balanced chemical reaction for the reduction of Cu(II) ions by zinc metal (12).

IX. Use a vacuum filter apparatus to filter your product of copper powder. Make sure the filter paper is seated by adding water from your wash bottle with the vacuum on. Pour the solution into the middle of the filter paper. (The copper powder is very fine and will be difficult to keep on the filter paper.) Rinse the test tube twice with a few mL of deionized water to remove the remaining copper powder from the test tube and pour these washes on the filter. Again be careful that the liquid being poured from the test tube doesn't wash the copper powder on the filter paper off the filter paper. Carefully wash the solid on the filter paper twice with 5 mL portions of acetone.

X. Obtain a 2-dram vial with a top. Prepare a label for this vial that lists the following: Cu, your name and desk number. Place the label on the vial, and then weigh the vial with cap on the analytical balance. Record the weight (5). Remove the top of the vial and transfer the copper powder to the vial. Place the vial (without the top) in an oven for ten minutes to dry. Remove the vial from the oven and allow it to cool.

XI. Place the top on the vial and weigh the vial on the analytical balance. Record the weight (4). Calculate the percent recovery of elemental copper (7).

PRE-LAB QUESTIONS

1. Why is it important to wear safety goggles in this experiment? What specific chemicals in this experiment can burn your skin or damage your eyesight if proper precautions aren't taken?

    _____

    _____

    _____

2. Why is concentrated nitric acid added to the copper wire in the hood?

    _____

3. What are the four types of chemical reactions copper and its compounds undergo in this experiment?

    _____

    _____

    _____

    _____

Experiment 20
Data Sheet

Name _____

Date _____

CHEMICAL REACTIONS OF COPPER

(1) Weight of copper wire pieces and weighing boat on analytical balance   _____ g

(2) Weight of empty weighing boat on analytical balance   _____ g

(3) Weight of copper wire pieces   _____ g

(4) Weight of vial with product on analytical balance   _____ g

(5) Weight of empty vial on analytical balance   _____ g

(6) Weight of product   _____ g

(7) Calculate the percent recovery of elemental copper.

_____ %

Balanced Equations for Reactions

(8)

(9)

(10)

(11)

(12)

POST-LAB QUESTIONS

1. A student reported a 110% recovery of elementary copper. What are two possible errors that would give a higher weight of elemental copper product compared to the initial weight of copper wire pieces?

    _____

    _____

2. Give the type of chemical reaction for each of the reactions in the present experiment as identified on the data sheet by the numbers:

    (8) _____

    (9) _____

    (10) _____

    (11) _____

    (12) _____

3. For the reactions in question 2 labeled "redox reactions" indicate what is oxidized and what is reduced in each reaction.

    _____

    _____

    _____

Organic Compounds in Three Dimensions

OBJECTIVE

To become familiar with organic molecules in three dimensions and to understand isomerism of simple organic molecules.

MODELS

Use one of the kits in the lab to make models of organic compounds.

FACTS TO KNOW

Over 11 million of the more than 13 million known compounds are carbon compounds, and a separate branch of chemistry, **organic chemistry,** is devoted to the study of them. Two important reasons why there are so many organic compounds are (1) the ability of thousands of carbon atoms to be linked in sequence with stable carbon–carbon bonds in a single molecule and (2) the occurrence of isomers. The basic geometrics for organic compounds are illustrated below.

Tetrahedron	**Trigonal Planar**	**Linear**

There are four classes of hydrocarbons: **alkanes,** which contain C$-$C single bonds with a tetrahedral arrangement around each carbon; **alkenes,** which contain one or more C=C double bonds with trigonal planar geometry around the carbon atoms with double bonds; **alkynes,** which contain one or more C $\equiv$ C triple bonds with linear geometry around the carbon atoms with triple bonds; and the **aromatics,** which consist of benzene, benzene derivatives, and fused benzene rings.

Two or more compounds with the same molecular formula but different arrangements of atoms are called **isomers.** Isomers differ in one or more physical or chemical properties such as boiling point, color, solubility, reactivity, and density. Several different types of isomerism are possible for organic compounds. These include structural isomers (straight-chain and branched-chain), *cis* and *trans* isomers, and optical isomers.

The simplest example of structural isomers is for C$_4$H$_{10}$ which can be either butane or methylpropane.

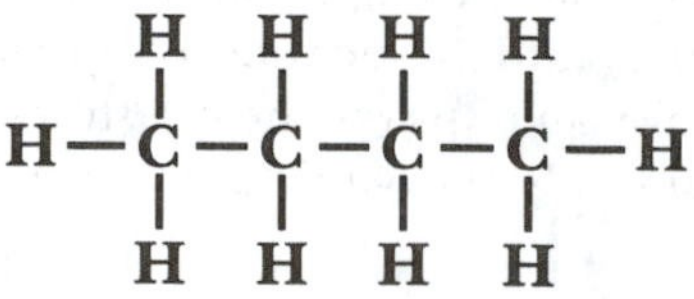

butane

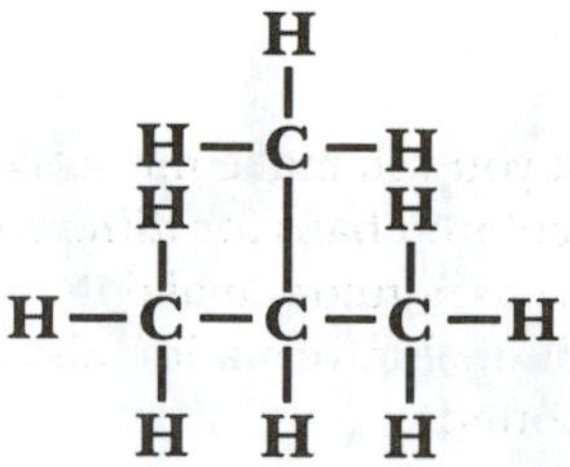

methylpropane

In the alkene series, the possibility of locating the double bond between two different carbon atoms add additional structural isomers. Ethene and propene have only one possible location of the double bond. However, the next alkene in the series, butene, has two possible locations for the double bond.

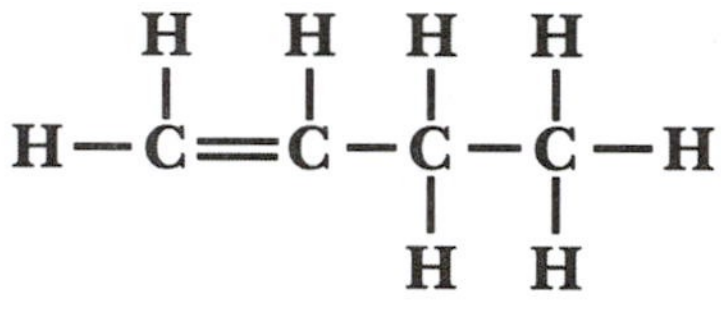

1–butene

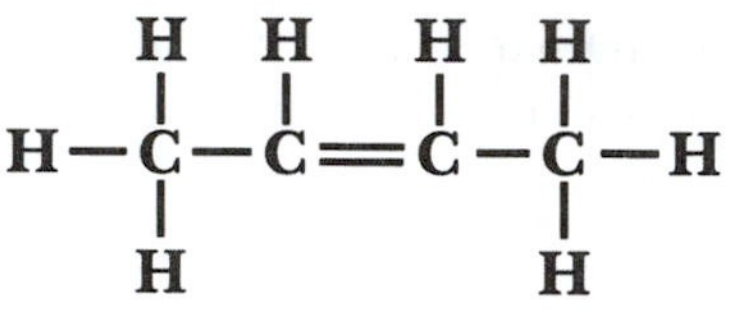

2–butene

An important difference between alkanes and alkenes is the degree of flexibility of the carbon–carbon bonds in the molecules. Rotation around single carbon–carbon bonds in alkanes occurs readily at room temperature, but the carbon–carbon double bond in alkenes is strong enough to prevent free rotation about the bond. This leads to the possibility of *cis-trans* isomers. *Cis-trans* isomerism in alkenes is only possible when both of the double-bond carbon atoms have two different groups.

If two methyl groups replace two hydrogen atoms, one on each carbon atom of ethene ($H_2C=CH_2$), the result is 2-butene, $CH_3CH=CHCH_3$. Experimental evidence confirms the existence of two compounds with the same set of bonds. The difference in the two compounds is in the location in space of the two methyl groups: the **cis** isomer has two methyl groups on the same side in the plane of the double bond and the **trans** isomer has two methyl groups on opposite sides of the double bond. The two arrangements are distinguished from each other by the prefixes *cis* indicating groups on the same side and *trans* indicating groups on opposite sides.

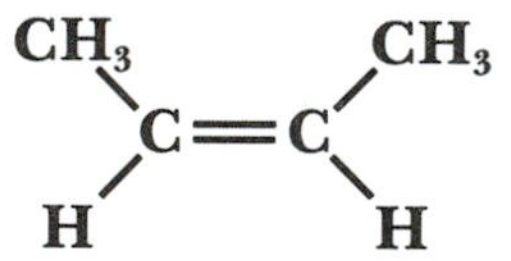

cis–2–butene

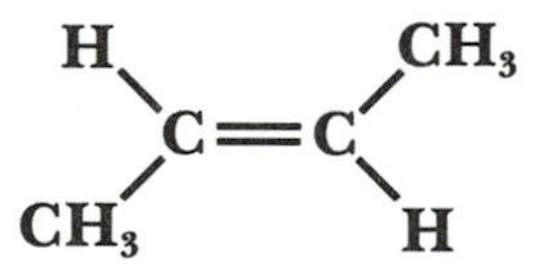

trans–2–butene

One class of optical isomers are found for compounds in which four different groups are attached to a given carbon atom as in CHXYZ. As can be seen from the images below, the resultant structures cannot be superimposed on each other no matter how they are turned about:

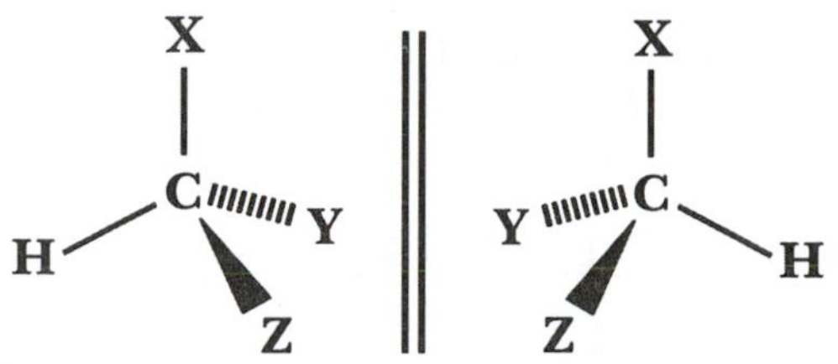

Mirror

PROCEDURE

In this experiment you will make models of organic compounds using plastic balls to represent the atoms (with different colored balls for different kinds of atoms) and plastic rods to represent the chemical bonds. Using these as a guide, make the models called for and answer the questions. Also consult your textbook for additional information about the different kinds of organic isomers and the rules for naming organic compounds.

PRE-LAB QUESTIONS

1. Why is C_4H_{10} the simplest alkane that can have structural isomers?

_____

_____

_____

2. Draw the *cis* and *trans* isomers for ClCH=CHCl. Why are *cis* and *trans* isomers not possible for $ClCH_2CH_2Cl$?

3. Draw the optical isomers for CHClBrF. Why are optical isomers not found for CH_2ClBr?

Experiment 21   **Name** _____
Data Sheet   **Date** _____

ORGANIC COMPOUNDS IN THREE DIMENSIONS

(1) Make a three-dimensional model of ethane. Are all six hydrogen atoms equivalent? Replace one hydrogen ball by a different colored plastic ball (representing Br). How many isomers of ethyl bromide (bromoethane) are possible? _____ Draw the structure(s) below.

(2) Make a model of propane. If you substitute one bromine atom for one hydrogen atom, how many isomers are possible? Show what they are and name them.

(3) How many isomers are possible for an alkane with the formula C_5H_{12}? Draw the structures and name them.

(4) Make a model of 1,2-dibromoethene. If *cis–trans* isomers are possible, make models of them. Draw the structures below and name them.

(5) Make a model of 2-methyl-1-butene. How does this differ from 2-methyl-2-butene? Draw their structures.

(6) How many butenes (C_4H_8) are possible? Draw their structures and name them.

(7) Make a model for the molecule CHClBrF. Can the atoms be arranged in more than one way in three dimensions? If so, what is the relationship of the different arrangements? (What is the relationship of your face to the face you look at in the mirror every morning?) Draw structures of any arrange-ments that are different.

(8) Make a model of 2-butyne. Are *cis–trans* isomers possible?

(9) An alcohol is classified according to the kind of carbon atom which bears the $-OH$ group. A carbon atom is classified as primary, secondary, or tertiary according to the number of alkyl groups attached to it. If the carbon atom which bears the $-OH$ group has no or one alkyl group attached to it, the alcohol is a *primary alcohol*. If two alkyl groups are attached, the alcohol is a *secondary* alcohol; if three alkyl groups are attached, the alcohol is a *tertiary* alcohol.

| primary alcohol | secondary alcohol | tertiary alcohol |

In general, alcohols undergo reactions which are dependent on the type of alcohol used. For example, primary alcohols are oxidized more readily than are secondary alcohols, and tertiary alcohols are oxidized only under severe conditions. Other reactions show similar trends.

Make models of primary, secondary, and tertiary alcohols that have the formula C_4H_9OH. Draw their structures below and name them.

POST-LAB QUESTIONS

1. Why does 2-butene have *cis* and *trans* isomers but 1-butene doesn't?

    _____

    _____

    _____

2. Draw the structural formula of 2-methyl-2-hexene.

3. Which of the following molecules can have optical isomers?

    (a) CH_2Cl_2 (b) $H_2NCH(CH_3)COOH$ (c) $ClCH(OH)CH_2Cl$

22 Rates of Chemical Reactions

OBJECTIVE

To measure the rate of a chemical reaction.

✓ To measure the effect of concentration upon the rate of a chemical reaction.

To demonstrate the effect of temperature on the rate of a chemical reaction.

To illustrate the method of obtaining a standard graph and of using it to determine the concentration of an unknown solution.

APPARATUS AND CHEMICALS

250-mL beakers (2)
50-mL burets (2)
150-mL beakers (4)
funnels (2)
50-mL volumetric flasks with stoppers (2)
buret stand and clamp

thermometer
solution #1 (4.28 g KIO_3 per liter)
solution #2 (0.852 g $NaHSO_3$, 4 mL concentrated H_2SO_4, and 50 mL of starch suspension per liter of solution)
ice
stirring rods (2)

SAFETY CONSIDERATIONS

 Caution should be used in handling solutions in this experiment. Solutions spilled on the skin should be washed off immediately with large quantities of water.

 Wear safety goggles. Solutions used in this experiment are harmful to the eyes.

FACTS TO KNOW

All chemical reactions go at definite rates that depend primarily upon the *nature of reactants,* their *concentration,* the *temperature,* the *surface area* of the reactants exposed, and the presence of *catalysts.* Many reactions, such as explosions of mixtures of hydrogen and oxygen, proceed so rapidly as to make accurate determination of their rates extremely difficult. Others, such as the rusting of iron, take place so slowly that rate determinations again are hard to make. However, there are many reactions, including the ones to be studied in this exercise, that occur at easily measurable rates.

Considering the theoretical submicroscopic structure of matter, a chemical reaction can only occur when the atoms, ions, or molecules of the reactants are in contact. Thus, the rate of a given reaction will depend on the frequency with which the reacting particles collide (Fig. 22-1a). Not all collisions produce a chemical reaction, however, because they do not meet the energy requirements for a reaction or, in some cases, molecules are not properly aligned for a reaction. In these cases, the particles merely rebound upon collision (Fig. 22-1b). However, any change in conditions that will increase the number of collisions between the particles should increase the rate of a reaction. Experimentally, this is found to be the case. For example, the following changes in conditions *increase* the rate of chemical reactions (Fig. 22-2):

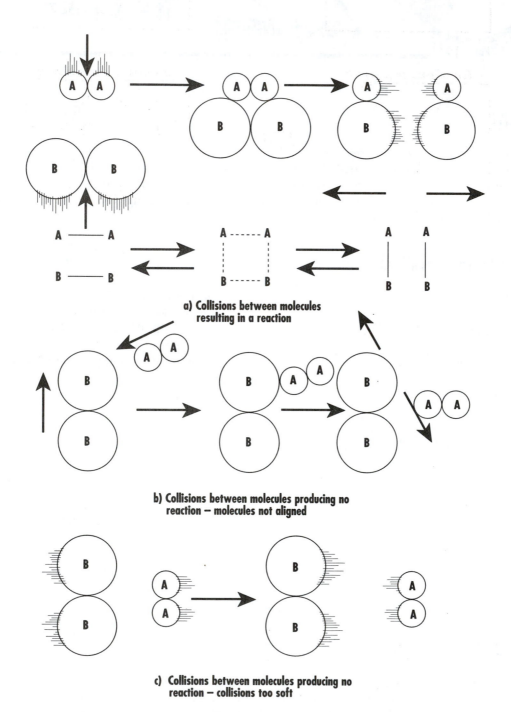

a) Collisions between molecules resulting in a reaction

b) Collisions between molecules producing no reaction — molecules not aligned

c) Collisions between molecules producing no reaction — collisions too soft

Figure 22-1. Molecule collisions and chemical reactions.

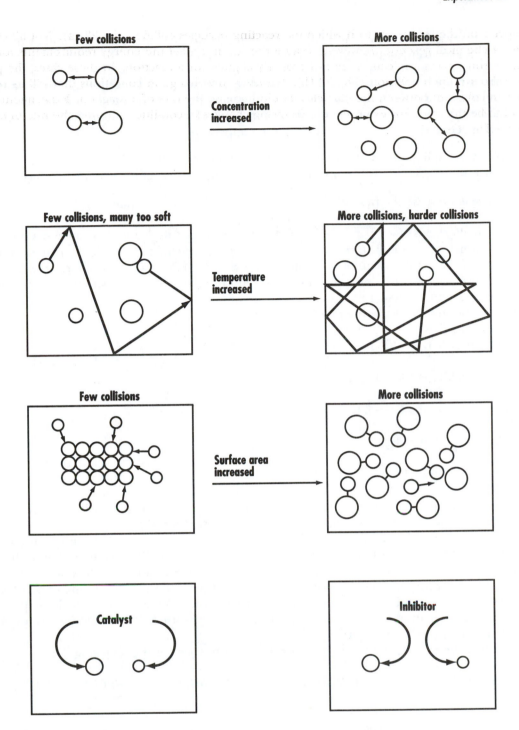

Figure 22-2. Factors affecting reaction rate.

a. An increase in the *concentrations* of the reactants. Since this increases the number of particles in any given volume, collisions will be more frequent.

b. An increase in the *temperature*. This makes the particles move faster; thus, they have more collisions in a given time. The collision rate is also increased because more molecules will have sufficient energy to overcome the energy barrier for reaction. Often, the rate of a chemical reaction doubles for each 10°C rise in temperature.

c. An increase in *surface* area of the particles. A lump of coal requires considerable time to burn, whereas coal dust can produce an explosion, because of its increased surface area.

d. *Catalysts* increase the rate of a reaction; inhibitors decrease the rate.

Although all of these factors may affect chemical reactions at the same time, an experiment can be designed in which all of the factors but one are the same in each trial. This is the case with the reactions in this experiment, a group of reactions known as the "Iodine Clock" reaction. In each trial, the concentration of one reactant differs from its concentration in other trials, whereas the temperature, sizes of particles, and catalyst are the same in every case.

The reactions for this experiment are the following:

2 $NaHSO_3 + H_2SO_4 \rightarrow H_2SO_3 + NaHSO_4$ (1)

#1 $KIO_3 + 3\,H_2SO_3 \rightarrow KI + 3\,H_2SO_4$ (2)

$KIO_3 + 3\,H_2SO_4 + 5\,KI \rightarrow 3\,K_2SO_4 + 3\,H_2O + 3\,I_2$ (3)

$I_2 + H_2SO_3 + H_2O \rightarrow H_2SO_4 + 2\,H^+ + 2\,I^-$ (4)

The more H_2HUD the longer the reaction

$I_2 + starch \rightarrow starch \cdot I_2$ (blue-black in color)   (5)

Reactions (1) through (4) continue until all of the $NaHSO_3$ is gone. Then the I_2 is no longer consumed by reaction (4) so I_2 reacts with starch to give a blue color (reaction (5)).

A standard graph will be made by reacting five solutions of five different concentrations, noting the time required for each reaction, and plotting the time of the reaction versus the concentration of the ingredient, which has been varied. The concentration of a solution of unknown concentration can be determined by matching the time required for its reaction with the corresponding concentration on the standard graph. In this experiment, a graph that is easier to read can be made by plotting the reciprocal of the measured time (1/time) rather than the time itself. Reciprocal time is proportional to the rate or the speed of this chemical reaction.

To illustrate how to plot and to use the standard graph, consider the following data:

SOLUTION	CONCENTRATION OF KIO_3 (M)	TIME (sec)	RECIPROCAL OF TIME (sec)$^{-1}$
A	0.001	55	0.018
B	0.002	39	0.026
C	0.003	29	0.034
D	0.004	22	0.046
E	0.005	20	0.050

The five points are plotted on the graph (Figure 22.3) by matching the concentration of each solution with its reciprocal time. To plot the point (A, 0.018), go up the A solution line until a value of 0.018 is reached. Mark a point and circle it. Plot the other 4 points in a similar manner. Draw the best straight line through your points by "averaging" the line between the points. In the graph shown, two data points are slightly above the line, two data points are slightly below the line, and one data point is on the line.

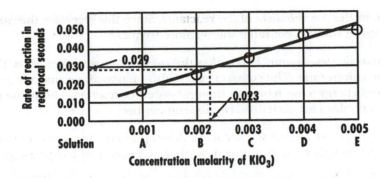

Figure 22-3. "Iodine Clock" reaction rate at various concentration of KIO_3.

To find the concentration of your unknown solution, locate its reciprocal time on the vertical axis, go horizontally across to the line that you have drawn on the graph, and then drop straight down to the horizontal (or concentration) axis. The concentration of your unknown is marked by the point at which this vertical line crosses the horizontal axis.

For example, suppose your unknown reacted in 34 seconds (reciprocal time is $1/34 = 0.029$). The concentration of your unknown is 0.0023, as shown by the dotted lines on the graph.

PROCEDURE

Numbers in parentheses refer to entry numbers on the data sheet.

 I. Clean two burets until water does not cling to the inside surface; rinse three times with distilled water. Place the burets on the buret stand (Fig. 22-4).

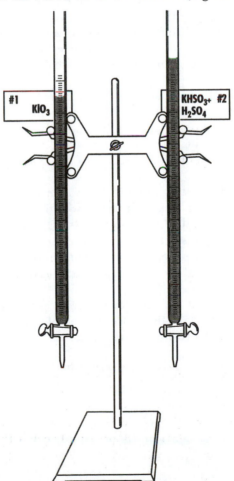

Figure 22-4. Arrangement of burets.

Clean, dry, and number two 250-mL beakers. Obtain about 100 mL of solutions #1 and #2. Solution #1 contains potassium iodate (KIO_3) and solution #2 contains sodium bisulfite ($NaHSO_3$), sulfuric acid (H_2SO_4), and starch.

Rinse the buret on the left twice with 5-mL portions of solution #1. Let the rinsings run out through the tip and discard them. Fill the buret above the zero mark and place the buret in the clamp on the stand. Remove the air bubbles in the tip as demonstrated by the instructor. (See Experiment 15 for detailed explanation of the use of a buret.) In a similar manner, prepare the buret on the right with solution #2.

II. Prepare solutions A through E *as listed on the data sheet* in the following manner. Into a 50-mL volumetric flask labelled #1, deliver directly from the buret the **exact quantity of solution #1** listed on the data sheet. Complete the filling of the flask to the 50-mL mark by adding distilled water. Shake well. Use another 50-mL volumetric flask and repeat the procedure with solution #2. In order to combine the two solutions quickly, transfer the 50 mL of solution #1 to a clean, dry 150-mL beaker. Place a clean stirring rod in this solution. Transfer the 50 mL of solution #2 to *another* clean, dry 150-mL beaker. Pour solution #2 into solution #1 and mark this as the starting time. Stir the solution for 2 or 3 seconds. Note the number of seconds required for the appearance of the blue color.

Repeat this procedure for each of the other solutions, B through E. The 50-mL volumetric flasks should be rinsed with water between each use. Repeat each run until you have confidence in the measurement.

III. Prepare the standard graph by plotting the concentration of KIO_3 on the horizontal axis and the reciprocal of the time on the vertical axis for solutions A through E. Draw the best straight line through your points.

IV. Repeat this procedure for solution F. Note that in this trial the concentration of solution #2 is changed (doubled) from the previous solutions, A through E. What does solution F indicate about the relationship between the amount of reactants and the speed of this reaction? For example, if you double the concentration of one reactant, what happens to the time of the reaction (1)? You should compare the reaction times of solutions B, D, and F.

V. Prepare solution G, which is similar to solution A except that you use cool water (about 15°C below room temperature) to fill the 50-mL volumetric flask to the mark. Compare this reaction time with that observed for solution A (2). Why is it important to use water from the same containers in all dilutions, A through E, rather than to use water from different containers (or tap water) for the first part of the experiment (3)?

VI. Rinse volumetric flask #1 with distilled water. Obtain a portion of the KIO_3 solution (#1), which will be your unknown, from your instructor. Complete the filling of this volumetric flask with distilled water to the 50-mL mark. Prepare 50 mL of solution #2 by combining 10 mL of the stock solution in the right hand buret and distilled water to the mark. Using beakers as before, pour solution #2 into solution #1, begin the timing, and stir for 2 or 3 seconds. Record the time for your unknown to react and calculate the reciprocal time.

Use the reciprocal of the time for this trial, as was described previously, and use the standard graph to determine the concentration of the unknown solution. Record this concentration in the blank provided on the data sheet.

PRE-LAB QUESTIONS

1. Probably the most dangerous accident which can occur in this experiment is that one of the solutions might be splashed into your eyes. What precaution must be taken to prevent this?

    _____

2. What is a catalyst?

    _____

3. If a certain reaction rate is 4.0 moles/min at a temperature of 20° C, what would be the approximate rate at 30° C?

    _____

4. Why must all air be expelled from the buret tip before accurate measurements can be made?

    _____

Experiment 22
Data Sheet

Name _____

Date _____

RATES OF CHEMICAL REACTIONS

Solution	mL of SOL. #1 ✓	mL of SOL. #2 ✓	Time for reaction	Reciprocal* of time	Molar concentration after mixing KIO_3	$NaHSO_3$
A	5 +45 H_2O	10 +40	_____	_____	0.0010 M	0.0008 M
B	10 +40 H_2O	10 +	_____	_____	0.0020	0.0008
C	15 + 35 H_2O	10 ⌐	_____	_____	0.0030	0.0008
D	20 +30 H_2O	10 ⌐	_____	_____	0.0040	0.0008
E	25 +25 H_2O	10 ⌐∨	_____	_____	0.0050	0.0008
F	20 +30 H_2O	20 + 30	_____	_____	0.0040	0.0016
G	5 +45 H_2O	10+40	_____	_____	0.0010	0.0008
Unknown		10	_____	_____	_____	0.0008

*Reciprocal of times mean to divide 1 by the time in seconds. That is, the reciprocal of 40 seconds is 1/40 sec = 0.025 sec^{-1}.

Plot of Reciprocal Time vs. Concentration of KIO$_3$ in the "Iodine Clock" Reaction

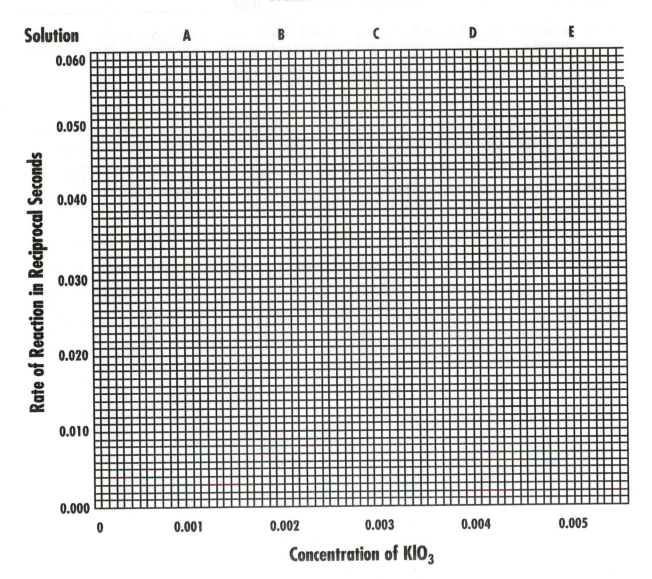

POST-LAB QUESTIONS

1. Refer to your results from solution F on the data sheet. On comparing this result from those of solutions B and D, what generalization can you make concerning the relationship between the rate of a reaction and concentration?

    _____

2. Compare results from solution G and solution A on the data sheet. Make a generalization concerning the effect temperature has on the rate.

    _____

3. Why is it important to use water from the same container to dilute all solutions rather than to use water from different containers for part of the experiment?

    _____

    _____

    _____

23   Analysis of Vitamin C

OBJECTIVE

To learn how to analyze food for vitamin C content, and to examine various sources of vitamin C for vitamin C content.

APPARATUS AND CHEMICALS

250-mL conical flask
400-mL beaker
100-mL beaker
50-mL buret with stand
glass stirring rod
analytical balance
balance
mortar and pestle

500 mg vitamin C tablets
5.00×10^{-2} M KIO_3 solution
0.60 M KI solution
1.0 M HCl solution
starch indicator solution (freshly prepared)
natural vitamin C sources(s)
cheesecloth

SAFETY CONSIDERATIONS

    Handle the glassware with caution to prevent breakage.

    When using a burner in the laboratory, be absolutely sure that loose hair and clothing do not come close to the flame. It is easy to forget this simple rule when concentrating on measurements.

    The solutions used in this experiment are not particularly caustic to the skin if washed off immediately, but the KIO_3 and HCl can both damage eye tissue. Wear your safety glasses at all times.

    Be sure the gas is turned off completely when you are through with it since in a closed laboratory a small gas leak can create an explosive mixture.
    Even though the substances being analyzed are foodstuffs, do not taste or eat any of the materials.

183

FACTS TO KNOW

Vitamins are complex organic molecules required in small amounts by the body in order to maintain health and well-being. Generally, the daily requirements of the various vitamins are very small quantities, but whenever these small quantities are not available, the body cannot function properly. For example, if a person's diet does not include a sufficient amount of vitamin C (the Food and Drug Administration recommends around 50 milligrams per day per person) a disease called scurvy results, which causes sores to develop in and around the mouth and gums. If enough vitamin A is not available, bones will not grow properly (among other problems) and this will lead to a malady called rickets.

Vitamin C is also known as ascorbic acid, and has the structure shown below.

$$
\begin{array}{c}
\text{HO} \\
 \\
\text{HO}
\end{array}
$$

In this experiment you will learn to analyze a sample of material for vitamin C content. The analysis makes use of the reducing qualities of the vitamin C. Both iodide (I^-) ions and iodate (IO_3^-) ions are added to a solution containing the vitamin C. The iodide and iodate ions react with each other to form elemental iodine (I_2) (Equation 1). The ascorbic acid (vitamin C, $C_6H_8O_6$) quickly reacts to reduce I_2 to I^-, destroying the vitamin C in the process (Equation 2). When all of the vitamin C has reacted, the next tiny amount of I_2 formed reacts with a small amount of starch, which has been added as an indicator, to produce a deep blue color (Equation 3). The iodate (as KIO_3 solution) is added with a buret, and the blue color of the resulting starch-iodine product signals the endpoint of the titration.

$$IO_3^- + 5\,I^- + 6\,H^+ \rightarrow 3\,I_2 + 3\,H_2O \tag{1}$$

$$I_2 + C_6H_8O_6 \rightarrow C_6H_6O_6{*} + 2\,H^+ + 2\,I^- \tag{2}$$

$$I_2 + \text{starch} \rightarrow (I_2 \cdot \text{starch}) \text{ (deep blue color)} \tag{3}$$

*oxidized product of vitamin C

PROCEDURE

Numbers in parentheses refer to entry numbers on the data sheet.

ANALYSIS OF COMMERCIAL VITAMIN C TABLETS Clean, rinse, and fill a buret with 5.0×10^{-2} M KIO_3 solution. Weigh a small piece of paper on an analytical balance (1). Then place a 500-mg vitamin C tablet on the paper and weigh the two together (2). What does the tablet actually weigh (3)?

Dissolve the tablet in about 150 mL of distilled water using a 250-mL conical flask. Crush the tablet carefully with a stirring rod to help it dissolve faster. Note that some insoluble material remains from the tablet. How does this relate to the weight of the tablet noted above?

Add to the flask 5 mL of 1.0 M HCl (this speeds up the reaction between the I^- and the IO_3^-), 10 mL of 0.60 M KI solution, and 2 drops of freshly prepared starch indicator solution.

Refer to Experiment 15 for titration techniques. Fill a buret with KIO_3 solution and take an initial buret reading (4). Carefully add KIO_3 solution from the buret until the solution takes on a permanent bluish-purple color. Record the final buret reading (5). What is the volume of the KIO_3 solution required for the titration (6)?

Calculate the *titre value* of the tablet (the titre value for a substance is the volume of titrant divided by the weight of the sample) (7). For example, if your tablet weighed 0.6012 g and required 22.45 mL of KIO_3 solution for titration, the titre value would be calculated as shown below:

$$22.45 \text{ mL}/0.6012 \text{ g} = 37.34 \text{ mL/g}$$

The titre value can be used as a measure of purity or for comparing the amount of vitamin C in different sources.

Repeat the analysis with a second tablet, and compare your titre values (8–14).

Weigh a third tablet and dissolve it in distilled water as before. Boil the solution for 15 minutes; cool it to room temperature; add the KI, HCl, and starch as before. Titrate to the blue endpoint (15–21). Is vitamin C affected appreciably by cooking (22)?

ANALYSIS OF NATURAL SOURCES OF VITAMIN C Instructions are given below for analyzing several natural products for vitamin content. Your instructor will assign particular products to be analyzed (23 and 29). (Different students may analyze different products so that more foods may be examined by the class.)

Be careful with these titrations. Only small amounts of titrant will be needed in each case.

1. *Citrus juices:* Most citrus juices contain pulp from the fruit, which makes the endpoint hard to see. Remove the pulp before using the juice by straining the juice through cheese cloth. Then measure out 100 mL of the juice (24), add 50 mL of distilled water, 5 mL of 1.0 M HCl, 10 mL of KI solution, and 4 drops of starch. Titrate carefully as before (25 and 26). Remember, *the color of the solution will be a composite of the juice color and the color of the indicator.* The endpoint will be indicated by a *color change*, but not necessarily to purple.

Juices one can use (and compare) include fresh and frozen orange juice, reconstituted lemon juice, vitamin C enriched fruit drinks such as Hi C or Tang, grapefruit juice, etc.

2. *Fruits and Vegetables:* Weigh out about 100 g of fruit or vegetable (30). Crush with a mortar and pestle to break down the cellular structure, then wash juice and pulp into a 400-mL beaker. Fill to the 150-mL mark and boil for 15 minutes. Cool, strain through cheese cloth, and analyze in the same fashion as for the fruit juices (31 and 32).

One can determine a value for the vitamin C content of the foodstuffs as follows. Assuming the tablets contained exactly 500 mg of vitamin C (The Food and Drug Administration has something to say about this!), one can determine the amount of vitamin C in the sample of food titrated using a simple ratio method. The volume of titrant needed for the food sample should be in the same ratio to the vitamin C in the food as the volume needed for the tablet is to the vitamin C in the tablet.

$$\frac{500 \text{ mg vitamin C}}{\text{vol. titrant needed for tablet}} = \frac{? \text{ mg vitamin C}}{\text{vol. titrant needed for food}}$$

EXAMPLE CALCULATION

Suppose your 500-mg tablet took 22.45 mL of titrant, and the sample of food required 3.45 mL of titrant. The vitamin C in the food sample would be

$$\frac{500 \text{ mg}}{22.45 \text{ mL}} = \frac{? \text{ mg (food)}}{3.45 \text{ mL}}$$

$$? \text{ mg (food)} = \frac{(500 \text{ mg}) \times (3.45 \text{ mL})}{22.45 \text{ mL}} = 76.8 \text{ mg (in food)}$$

Determine the vitamin C content of each of the foods analyzed (28 and 34).

PRE-LAB QUESTIONS

1.   What precaution should be taken if any of the solutions in this experiment are spilled on your skin?

   _____

2.   Why should you not taste any substance used in this experiment?

   _____

3.   What substance actually reacts with the vitamin C in this reaction and what is the vitamin C converted into?

   _____

   _____

4.   In this reaction, a titre value is determined. What is a titer value?

   _____

   _____

Experiment 23
Data Sheet

DETERMINATION OF VITAMIN C

	Tablet 1	Tablet 2	Tablet 3 (boiled)
Weight of paper (g)	(1)_____	(8)_____	(15)_____
Paper + tablet (g)	(2)_____	(9)_____	(16)_____
Weight of tablet (g)	(3)_____	(10)_____	(17)_____
Initial buret reading (mL)	(4)_____	(11)_____	(18)_____
Final buret reading (mL)	(5)_____	(12)_____	(19)_____
Volume of KIO_3 (mL)	(6)_____	(13)_____	(20)_____
Titre value	(7)_____	(14)_____	(21)_____

(22) _____

_____

	Natural Product 1 (citrus juice)	Natural Product 2 (fruit or vegetable)
Product	(23)_____	(29)_____
Volume (mL) or wt. (g)	(24)_____	(30)_____
Initial buret reading (mL)	(25)_____	(31)_____
Final buret reading (mL)	(26)_____	(32)_____
Volume of KIO_3 (mL)	(27)_____	(33)_____
Vitamin C content	(28)_____	(34)_____

POST-LAB QUESTIONS

1. When weighing the vitamin C tablet in this experiment students frequently find that it weighs more than the stated value. For example, a 500 mg tablet might weigh more than 500 mg. Explain.

    _____

    _____

    _____

2. Often when dissolving vitamin C tablets some of the tablet does not dissolve. Does this mean that some vitamin C cannot be dissolved or is there another explanation?

    _____

    _____

    _____

3. The instruction for fruits and vegetables in this experiment tell you to boil the crushed material for 15 minutes to extract vitamin C from the fibers. Can this boiling affect the amount of vitamin C which you determine in your analysis? How?

    _____

    _____

    _____

24

The Preparation of Sodium Hydrogen Carbonate (NaHCO₃) and Sodium Carbonate (Na₂CO₃)

OBJECTIVE

To prepare a sample of sodium hydrogen carbonate ($NaHCO_3$) by the reactions used for its synthesis in the Solvay process.

To prepare a sample of sodium carbonate from sodium bicarbonate.

APPARATUS AND CHEMICALS

125-mL conical flask
funnel
tongs of tweezers
filter paper
250-mL beaker
100-mL graduated cylinder
vacuum filtration apparatus
watch glass

ring stand
test tube
litmus paper
phenolphthalein indicator solution
concentrated ammonium hydroxide (NH_4OH)
sodium chloride
dry ice
clamp

SAFETY CONSIDERATIONS

Concentrated ammonium hydroxide is a saturated solution of ammonia gas. Since ammonia, a strong bronchial irritant, is released when removing the cap from a concentrated ammonium hydroxide bottle, use concentrated ammonium hydroxide in the hood. If you are overcome by fumes, go to fresh air.

The temperature of dry ice is $-78°C$. It should not be handled with bare hands or fingers as it is capable of causing severe damage to the flesh.

FACTS TO KNOW

Sodium hydrogen carbonate, $NaHCO_3$, also called sodium bicarbonate, is made and used on a very large scale.

It is found in virtually every kitchen in the country (the orange box with the arm and hammer) where it is used in many recipes, for cleaning up spills, as a deodorizing agent in refrigerators, as a scouring powder, and sometimes as a toothpaste. It is used as an antacid as well as in the manufacture of other compounds such as sodium carbonate, Na_2CO_3. Sodium carbonate is used in the manufacture

189

of glass, cleaning powders, in the treatment of hard waters, and as a cheap, convenient industrial base. The reaction used here is the one which forms the basis of the Solvay process. When sodium chloride is dissolved in a concentrated solution of ammonia, the resultant solution absorbs carbon dioxide readily by the reaction

$$NH_3 + H_2O + CO_2 \rightarrow NH_4^+ + HCO_3^-$$

The ammonia is a base and the carbon dioxide is an acid in this neutralization reaction. While ammonium hydrogen carbonate is very soluble in such a solution, sodium hydrogen carbonate is not. It precipitates from such a solution as fine white crystals.

$$Na^+ + HCO_3^- \rightarrow NaHCO_3\downarrow$$

The precipitate is separated from the solution by filtration to give reasonably pure sodium hydrogen carbonate after it is dried at 100°C.

When heated to about 270°C, sodium hydrogen carbonate loses both water and carbon dioxide to give sodium carbonate:

$$2\ NaCHO_3 \underset{270°C}{\overset{\Delta}{\rightarrow}} Na_2CO_3 + H_2O + CO_2$$

Sodium carbonate is a more strongly basic material than sodium hydrogen carbonate because of the reaction (in water)

$$CO_3^{2-} + H_2O = HCO_3^- + OH^-$$

Solutions of sodium carbonate in water are basic to indicators such as phenolphthalein.

PROCEDURE

Numbers in parentheses refer to entry numbers on the data sheet.

I. PREPARATION OF SODIUM BICARBONATE   Transfer, under the hood, 60 mL of concentrated ammonium hydroxide into a 125-mL conical flask. Add about 13 g of sodium chloride in small portions to this, swirling the flask after each addition, until no more sodium chloride will dissolve. When the solution is saturated with sodium chloride (and contains crystals of salt which do not dissolve), filter the solution through a funnel containing filter paper and into a 250-mL beaker. *USE THE HOOD.* (If care is taken, it may be possible to decant the clear saturated solution into the beaker, making certain that no crystals of sodium chloride are carried along with the saturated solution.) Add about 50 g of dry ice in the form of several small pieces to this solution, using either tweezers or tongs to handle the dry ice. As the carbon dioxide saturates the solution, a white precipitate of sodium hydrogen carbonate forms. Wait until all of the dry ice has reacted or sublimed.

While the dry ice is reacting, set up a vacuum filtration apparatus with a Büchner funnel and filter paper (see Fig. 19-1).

After copious precipitate has formed, swirl to create a suspension and rapidly pour the suspension onto the filter paper in the Büchner funnel. Allow the vacuum to pull most of the water out of the precipitate. Dry the solid on a weighed watch glass placed over a beaker of gently boiling water. Heat the solid in this way until it no longer has the odor of ammonia. If the ammonia is not all removed at this point, the resulting solutions prepared from the solid will be more basic than normal. Weigh the dried precipitate (1).

II. PREPARATION OF SODIUM CARBONATE   Take a 1-g sample of your sodium bicarbonate (sodium hydrogen carbonate) and place it in a large Pyrex test tube. Heat it in a laboratory burner flame, at first rather gently and then in successively hotter parts of the flame. When there is no longer any evidence of reaction, set the test tube aside and allow it to cool for at least 10 minutes. The size of the solid material can be observed to shrink as the carbon dioxide and water are expelled.

III. PROPERTIES OF THE PRODUCTS   Take a small quantity of the sodium hydrogen carbonate you have prepared and dissolve it in 5 mL of distilled water. Test the solution with litmus paper (2). Add a few drops of phenolphthalein indicator solution and record the result (3). Finally, add a few drops of dilute sulfuric acid solution. Record the observation (4).

Take the cooled test tube containing the sodium carbonate you have prepared and dissolve approximately one half of it in about 5 mL of water. Test the solution successively with litmus paper (2), phenolphthalein indicator solution (3), and dilute sulfuric acid solution (4). Record your results. The sodium bicarbonate preparation will be more basic than normal if all the ammonia is not removed from it on the steam bath.

IV. PERCENTAGE YIELD   Calculate the percentage yield as follows (5 and 6):

$$\% \text{ yield} = \frac{\text{Weight of product (1)}}{\text{Theoretical yield of NaHCO}_3} \times 100\%$$

EXAMPLE CALCULATION OF THEORETICAL YIELD

13.0 g of NaCl used to prepare $NaHCO_3$
Formula weight of NaCl = 58.5
Formula weight of $NaHCO_3$ = 84.0

Theoretical yield of $NaHCO_3$ = 13.0 g NaCl $\times \dfrac{84.0 \text{ g NaHCO}_3}{58.5 \text{ g NaCl}}$

$\qquad\qquad = 18.7 \text{ g NaHCO}_3$

Transfer the remaining sodium hydrogen carbonate from Part I to a container; prepare a label for the container with your name, name of product, actual yield in grams, and percentage yield; and turn in the labelled container to your instructor.

PRE-LAB QUESTIONS

1. If you spilled concentrated ammonium hydroxide on your hands, what should be your immediate response?

   _____

   _____

   _____

2. List at least one pharmaceutical use of sodium bicarbonate.

   _____

   _____

   _____

3. Why should concentrated ammonium hydroxide be used in the hood?

   _____

   _____

   _____

4. In your initial procedure, what would be the effect of accidentally pouring off some of the solid sodium chloride as you decanted the liquid?

   _____

   _____

   _____

Experiment 24
Data Sheet

Name _____

Date _____

PREPARATION OF SODIUM HYDROGEN CARBONATE AND SODIUM CARBONATE

Record the weight of the dried sodium hydrogen carbonate you have prepared.

Weight of watch glass + $NaHCO_3$ _____ g

Weight of watch glass   _____ g

(1) Weight of $NaHCO_3$ _____ g

REACTIONS OF SOLUTIONS OF $NaHCO_3$ AND Na_2CO_3

REAGENT	$NaHCO_3$	Na_2CO_3
(2) Litmus paper	_____	_____
(3) Phenolphthalein	_____	_____
(4) Dilute H_2SO_4	_____	_____

CALCULATIONS

(5) Theoretical yield of $NaHCO_3$ _____ g

(6) Percentage yield of $NaHCO_3$ (yields are fairly low)   _____ %

POST-LAB QUESTIONS

1. List some factors which would cause your yield to be low.

   _____

   _____

   _____

2. What chemicals are produced during heating or applying acid to the sodium bicarbonate?

   _____

   _____

   _____

3. Does your mother keep sodium bicarbonate in the kitchen? If so, what does she call it?

   _____

   _____

   _____

4. Suppose you used 25.0 g of sodium chloride to saturate a solution of concentrated ammonium hydroxide followed by treatment with dry ice and produced 15 g of sodium bicarbonate. What would be the overall yield? (Show calculations)

25 The Preparation of an Aluminum Compound from Aluminum Cans

OBJECTIVE

To show how chemical processes can be used to transform scrap material into a useful compound—a recycling process.

APPARATUS AND CHEMICALS

250-mL beaker
Büchner funnel
conical funnel
150-mL beaker
labels
graduated cylinder
laboratory burner
vacuum filter flask
mortar and pestle
melting point tube
rubber band
sample bottle

alkali resistant filter paper
    (11 cm) (Whatman #54, hardened)
Büchner funnel filter paper
regular filter paper (11 cm)
aluminum scrap
potassium hydroxide solution
    (80 g/liter)
distilled water
9 M sulfuric acid
ethanol
congo red test paper

SAFETY CONSIDERATIONS

Hydrogen gas is produced in the reaction of aluminum with KOH so the reaction should be carried out in the hood. Hydrogen-oxygen mixtures explode in the presence of a spark or a flame.

Both strong base and strong acid are used in this experiment. Wear eye protection and avoid skin contact with the KOH and 9 M H_2SO_4 solutions. Both are corrosive liquids that can cause serious eye and skin damage. Denatured ethanol is a poison.

FACTS TO KNOW

Recycling of waste materials has become a primary way of doing business in the past decade. Once considered a "politically correct" procedure that was environmentally friendly but which frequently was more expensive than using virgin raw materials for production of new materials, recycling has become a necessity in many instances and a cheap source of raw materials for the production of new products. A profitable industry has developed which uses only discarded plastic gallon milk cartons as raw materials to make a variety of useful plastic products, as diverse as park benches and gear wheels in motors. Whereas only a few years ago it was difficult to even dispose of old newspapers because it was so expensive to recycle them, it is now not uncommon for stacks of newspapers to be stolen from collection points because new recycling technology has made them so valuable as raw materials.

Many people have the mistaken idea that recycling means turning discarded materials into newly manufactured pieces of the same materials—that is, discarded aluminum cans become new aluminum cans, milk cartons become new milk cartons, etc. As the example above of milk cartons becoming park benches illustrates, this is not the case at all, although it does occur to some extent. Some aluminum cans are recycled to produce new aluminum cans, but many become other products containing aluminum. Some of these products do not even resemble aluminum, as the products from this experiment will clearly illustrate.

Chemical reactions can provide the basis for the transformation of waste materials into new compounds having a completely different appearance and greater utility than the starting materials. The widespread use of aluminum for beverage cans has led to the realization that such cans are much more resistant than steel cans to disintegration by corrosion. This has caused increased concern over their ultimate disposition. The present experiment presents methods for synthesizing a useful aluminum compound from such scrap material. Aluminum dissolves rapidly in a hot aqueous solution of potassium hydroxide. The equation for this reaction is

$$2\,Al + 2\,KOH + 6\,H_2O \rightarrow 2\,K^+ + 2\,Al(OH)_4^- + 3\,H_2\uparrow$$
$$\text{(In solution)} \qquad \text{Gas}$$

In the course of this reaction the concentration of hydroxide ion in the solution is reduced considerably. After the aluminum has been dissolved, it can be easily transformed into a number of different compounds which are either valuable in their own right or useful as intermediates in the preparation of still other aluminum compounds.

Potassium aluminum sulfate ($KAl(SO_4)_2 \cdot 12H_2O$), or "alum," is an aluminum compound which is widely used in the dyeing of fabrics, the tanning of leather, in the manufacture of pickles, and for a host of other materials. One of its most important uses is in the water treatment industry, as a part of the clarification process, whereby suspended solids in water from a raw water source (muddy water from a river, for example) is clarified (made clear) as the first step in purifying it for use as potable water (drinking water) that comes out of the tap. In the dyeing industry, alum serves as a mordant in dyeing cloth. A mordant contains metal ions that bind dyes to the fabric. The ancient Egyptians, Greeks, and Romans used alum for this purpose, and this application is still one of the important uses of alum.

"Alum" is a generic term that refers to double salts with the general formula, $M_2SO_4 \cdot M'_2(SO_4)_3 \cdot 24H_2O$ in which M is commonly Na^+, K^+, NH_4^+, and M′ is Al^{3+}, Cr^{3+}, Fe^{3+}, or Co^{3+}. Since the most common alum is potassium aluminum sulfate, the word "alum" is most often used to refer to this compound. Other alums use the name of the trivalent ion along with the word "alum." For example, the alum with Cr^{3+} in place of Al^{3+} is called chromium alum or chrome alum.

All of the alums form well-shaped crystals when they crystallize from solution. An optional part of this experiment is to use the prepared alum to grow large crystals.

PROCEDURE

Weigh out 1 g of aluminum scrap, cut it into very small pieces, and put these in a 250-mL beaker. Add 50 mL of the potassium hydroxide solution. BE CAREFUL! DO NOT SPLATTER THE SOLUTION. At this point hydrogen will be evolved copiously, so the experiment must be carried out in well circulating air such as under a hood. Place a beaker on a hot plate in the hood. The temperature setting on the hot plate should be as low as possible to keep the solution from getting too warm. After about 10 minutes the aluminum should be mostly dissolved, except for small amounts of impurities. Heat until the fizzing around each piece stops.

You may need to add a small amount of water from your squeeze bottle to this reaction mixture occasionally to compensate for water lost by evaporation during the heating. Do not allow the total volume of the solution to *increase* over the original 50 mL through these additions, but maintain the level at around 50 mL throughout the heating. When all of the aluminum has dissolved, the fizzing around the small pieces will have ceased. At this point, there will be a good deal of solid remaining in the beaker. This solid is the plastic label that covered the can, along with residue of the glue used to adhere the label to the can.

Allow the solution to cool for about 5–8 minutes, then *slowly and carefully* add 20 mL of 9 M H_2SO_4. This neutralizes the excess KOH and acidifies the solution. The alum is made during this addition, and you will see it appear as a white, waxy-looking solid during the addition of the acid.

$$2 K^+ + 2 Al(OH)_4^- + H_2SO_4 \rightarrow 2 K^+ + 2 Al(OH)_3 \downarrow + 2 H_2O + SO_4^{2-}$$

$$2 Al(OH)_3 + 3 H_2SO_4 \rightarrow 2 Al^{3+} + 3 SO_4^{2-} + 3 H_2O$$

After all of the acid has been added, heat the beaker until the solution just begins to boil, then quickly filter the solution through qualitative filter paper (Fig. 25-1), collecting the filtrate in a clean beaker. You should obtain a clear, colorless solution and the remnants of the label will be left on the filter paper. The solution is filtered *after* it is acidified because hot alkaline solutions quickly dissolve paper and this makes filtration of the hot basic solution more difficult. It can be done, however, if alkali-resistant filter paper is used, but this is more expensive.

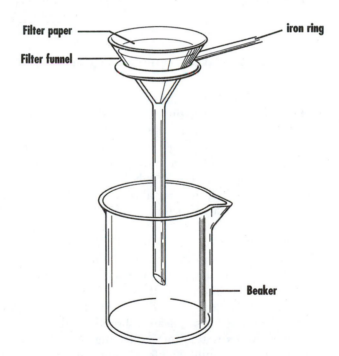

Figure 25-1. Filtration set-up.

Allow the host solution to cool for about 10 minutes, then place the beaker in an ice bath for 20 minutes. At the end of this time, alum crystals should be present in the beaker. These crystals are then collected with the vacuum filter (Fig. 25-2). Solutions of alum have a pronounced tendency to supercool and frequently crystals do not form as a result. This experiment provides an excellent opportunity to see various methods of inducing crystallization in a supercooled solution (scratching the bottom of the beaker with a glass stirring rod, addition of a seed crystal, etc. Your instructor may choose to demonstrate these echniques with some of the student-prep solutions that have supercooled.

If it appears you will have time to isolate a second crop of alum crystals, pour the solution from the filter flask into a 250-mL beaker before washing the crystals with the alcohol-water mixture.

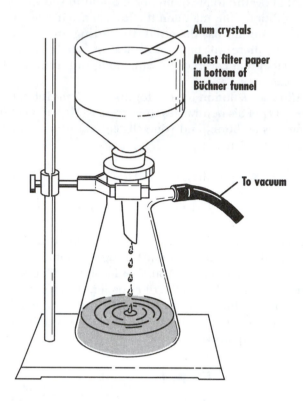

Figure 25-2. Vacuum filtration apparatus for filtration of alum.

Wash the alum crystals on the Büchner filter paper with 20 mL (5 mL at a time) of a 50/50 ethanol–water mixture (in which alum is not very soluble), apply vacuum after each wash to pull the wash liquid through the filter, and then leave the vacuum on to pull air through the filter paper until the crystals are dry. While drying, cut up the crystals with a spatula. Weigh your crystals. This is the "Grams of product" for your label. If time permits, obtain a second crop of alum crystals by evaporating the solution you poured into the 250-mL beaker to one-half its original volume. Cool the solution in an ice bath and collect the second crop of crystals as before.

MELTING POINT An accurate value of the melting point of the alum can be obtained only if the crystals are very dry. This may require overnight or until-next-lab drying time. If the melting point is to be taken, obtain a melting point tube or a capillary tube and small rubber band from the instructor. If the tube is not already closed at one end, heat one end of the capillary tube in the edge of the flame of a gas burner until the end completely closes. The end that is heated should be above the open end during the heating process. This prevents water from collecting in the closed end of the capillary. Rotate the tube between your thumb and forefinger during heating to prevent it from bending.

Pulverize about 1 g of the sample with a clean mortar and pestle. Carefully push the open end of the capillary tube into the powdered sample, forcing a portion of the sample into the tube. Invert the tube and flick the tube with your finger, rasp the top with a file to vibrate the sample into the closed end, or drop the capillary down a long glass tube onto the table. Repeat until the capillary contains sample to a depth of about ½ cm. Do not press too much sample into the capillary tube during each step. It will pack tightly near the top and then cannot be shaken to the bottom of the tube.

Place a small rubber band 3 cm to 5 cm above the bulb of the lower end of a thermometer. Insert the capillary tube under the rubber band with the closed end and sample near the bulb. Place the thermometer into a cork or rubber stopper so that the stopper may be secured in a buret clamp. Place the thermometer and capillary tube in a 250-mL beaker containing enough water to cover the bulb and portion of capillary containing the sample. Be sure that the open end of the capillary is above the water level. Figure 25-3 shows the arrangement of equipment for melting point determination.

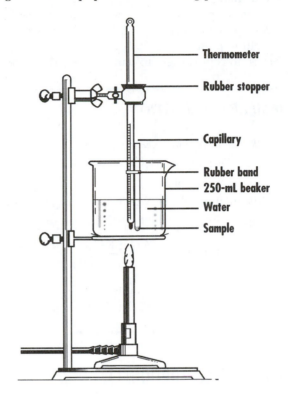

Figure 25-3. Vacuum filtration apparatus for filtration of alum.

Heat the water slowly—try for a rate of increase about 1° per 20 seconds. Observe the solid in the capillary very carefully. At the moment that the solid melts, note the temperature. This is the melting point of the solid. Record this value on your label.

ACIDITY   Carefully weigh 1.00 g of your sample into a small beaker. Add 25 mL of distilled water and stir with a glass stirring rod until the solid has dissolved. Dip a piece of congo red test paper into this solution and tape the paper to your sample bottle.

CALCULATIONS AND REPORT   The reactions for the preparation of alum are

$$2\,Al + 2\,K^+ + 2\,OH^- + 6\,H_2O \rightarrow 2\,K^+ + 2\,Al(OH)_4^- + 3\,H_2 \uparrow$$

$$2\,K^+ + 2\,Al(OH)_4^- + 4\,H_2SO_4 \rightarrow 2\,K^+ + 2\,Al^{3-} + 4\,SO_4^{2-} + 8\,H_2O$$

$$K^+ + Al^{3+} + 2\,SO_2^{2-} + 12\,H_2O \rightarrow KAl(SO_4)_2 \cdot 12H_2O$$

From the balanced equations it can be seen that 1 gram atom of aluminum yields 1 gram formula weight of alum. Calculate the theoretical weight of alum obtainable from 1 g of aluminum.

$$\text{Theoretical weight of alum} =$$

$$\text{wt. of aluminum} \times \frac{\text{gram formula wt. of } KAl(SO_4)_2 \cdot 12H_2O}{\text{atomic wt. of } Al}$$

Then calculate *your* percent yield:

$$\% \text{ yield} = \frac{\text{weight of alum obtained}}{\text{theoretical weight of alum from 1 g of Al}} \times 100$$

Turn in your remaining sample in a container prescribed by the instructor. Include the following on the label:

Your name
Grams of product
Your percent yield
Melting point

There is no report sheet to hand in with this experiment.

EXAMPLE CALCULATION

1.00 g of aluminum taken
14.2 g KAl(SO₄)₂·12H₂O (recovered)

1.00 g of aluminum taken
14.2 g KAl(SO$_4$)$_2$·12H$_2$O (recovered)
Atomic weight of Al = 27.0
Molecular weight of KAl(SO$_4$)$_2$·12H$_2$O = 474
Weight of KAl(SO$_4$)$_2$·12H$_2$O (theoretical) =

$$1.00 \text{ g Al} \times \frac{474 \text{ g KAl(SO}_4)_2 \cdot 12\text{H}_2\text{O}}{27.0 \text{ g Al}} = 17.5 \text{ g}$$

$$\% \text{ yield} = \frac{14.2 \text{ g KAl(SO}_4)_2 \cdot 12\text{H}_2\text{O (recovered)}}{17.5 \text{ g KAl(SO}_4)_2 \cdot 12\text{H}_2\text{O (theory)}} \times 100 = 81.0\%$$

GROWING LARGE CRYSTALS (OPTIONAL)

After your lab instructor has checked your product and returned it to you, combine your product with that of other students until the total amount is about 20 g to 30 g.

Weigh out 10 g of alum, and place it in a 250-mL beaker. Add 70 mL of water and heat the beaker to 50°C to 60°C with stirring until the solution is clear. In case the solution does not become clear, let it stand for a few minutes until the sediment has settled out and then decant off the clear solution.

Tie a piece of thread to a glass rod or wooden splint so that the thread will extend no more than one-half inch below the surface of the solution when the glass rod or splint is placed across the top of the beaker. Smear stopcock grease on the part of thread which will be *above* the solution to keep the solution from creeping up the thread. Put the rod or splint in place on top of the beaker, cover the solution with a watch glass, and place the beaker in a safe storage place where the beaker will not be jarred or moved. Crystals suitable for use as "seeds" should have formed by the next laboratory period.

At the beginning of the next laboratory period, check to see if any crystals have formed on the thread. If they have, crush off all but the best one and use it as the seed crystal. If the crystals are on the bottom of the beaker, decant the liquid from the crystals into another clean beaker. Inspect the crystals and pick out one which has clearly-defined faces. Tie a thread to this seed crystal and suspend it from a rod or splint.

Suspend the seed crystal in a *freshly* prepared solution of 10 g of alum in 70 mL of water. This solution *must* be allowed to cool to room temperature before the seed crystal is introduced.

Repeat the procedure of changing solutions every few days until you have a crystal the size you want to keep. To preserve it and avoid its conversion to a white powder (by loss of water) cover the crystal with a clear plastic spray available from your lab instructor.

Mixed crystals can be grown by substituting a solution of chrome alum, KCr(SO$_4$)$_2 \cdot$ 12H$_2$O, for the *second* alum growing solution. Prepare your seed crystal as described above. Then prepare a chrome alum solution by adding 17 g of chrome alum to 25 mL of water. Also prepare an alum solution by dissolving 8 g of alum in 56 mL of water. Slowly add the chrome alum solution to the alum solution while holding the alum solution up to the light. Stop adding the chrome alum solution when you can just see through the solution. Suspend your seed crystal in the solution as before. For subsequent growing stages, use ordinary alum solutions. This will give you a purple crystal inside a clear crystal.

When changing your growing crystal to a fresh solution, be sure the solution is cool and at the proper concentration. Otherwise, your crystal may dissolve. You can check for this by observing the portion of the solution near the suspended crystal. If the solution is less than saturated, the portion of the solution in contact with the crystal will dissolve some of the crystal. This will increase the density of the solution around the crystal and set up a density current as the heavier solution near the crystal moves to the bottom. If this happens, remove the crystal and allow the solution to cool further or if cool, add more alum to saturate the solution before putting the growing crystal back into the solution.

PRE-LAB QUESTIONS

1. What is meant by "recycling" in the context of this experiment?

   _____

   _____

   _____

2. What are the common constituents of the formulas of the group of compounds collectively known as "alums?"

   _____

   _____

   _____

3. Why is the solution acidified before it is filtered to remove the solid remaining after the aluminum has dissolved in the KOH solution?

   _____

   _____

   _____

4. The formula for alum contains SO_4^{2-} (sulfate ions). What is the source of the sulfate ions in this procedure?

   _____

   _____

   _____

5. List 3 commercial uses for alum.

   _____

   _____

   _____

POST-LAB QUESTIONS

1. It is possible to prepare almost 18 grams of alum beginning with only one gram of aluminum. Explain how this does not violate the law of conservation of mass.

    _____

    _____

    _____

2. Why was the alum product washed with a 50% alcohol solution, and what is the function of the alcohol in this solution?

    _____

    _____

    _____

3. Why was it important to not increase the volume of the reaction solution by adding too much water while the aluminum was dissolving?

    _____

    _____

    _____

26 The Comparison of Saturated, Unsaturated, and Aromatic Hydrocarbons

OBJECTIVE

To learn some of the characteristic differences in the reactions of saturated, unsaturated, and aromatic hydrocarbons.

APPARATUS AND CHEMICALS

watch glasses (4)
test tubes (4)
Bunsen burner
copper wire
petroleum ether
hexane
cyclohexane

toluene
chlorobenzene
sulfuric acid
bromine in cyclohexane
potassium permanganate solution
ultraviolet lamp

SAFETY CONSIDERATIONS

The Bunsen burner is used in this experiment. Care should be exercised in igniting and using the burner. Do not forget to turn the gas off after use. The solvents used in this experiment are flammable. Avoid using them near an open flame.

Concentrated sulfuric acid causes severe burns to the skin and eyes. This acid will also destroy clothing. Handle sulfuric acid with great care. Bromine and toluene are toxic and should be handled in adequate ventilation.

Potassium permanganate is a poison and a strong oxidizing agent. Handle it with care.

FACTS TO KNOW

In this experiment we will compare the properties of a saturated hydrocarbon (hexane) with those of an unsaturated hydrocarbon (cyclohexene) and two aromatic compounds (toluene and chlorobenzene).

A saturated hydrocarbon is a compound such as *n*-hexane

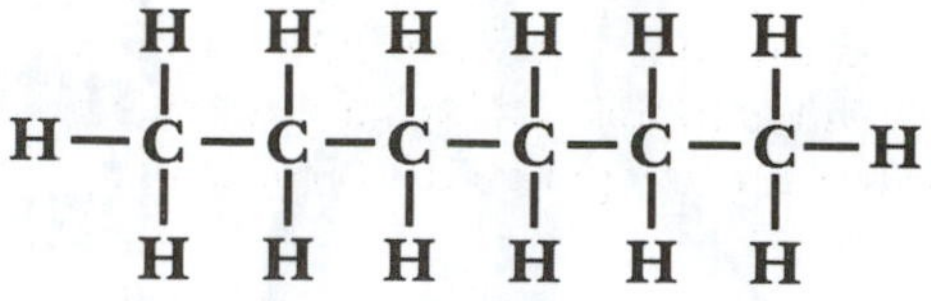

$$\begin{array}{cccccccccccc}
 & H & & H & & H & & H & & H & & H \\
 & | & & | & & | & & | & & | & & | \\
H- & C & - & C & - & C & - & C & - & C & - & C & -H \\
 & | & & | & & | & & | & & | & & | \\
 & H & & H & & H & & H & & H & & H
\end{array}$$

n–hexane

in which each carbon has four single bonds, joining it to four other atoms (carbon or hydrogen) which surround it tetrahedrally. An unsaturated hydrocarbon is a compound containing carbon and hydrogen and one or more double bonds between adjacent carbon atoms. Examples of unsaturated hydrocarbons include compounds such as ethylene and cyclohexene:

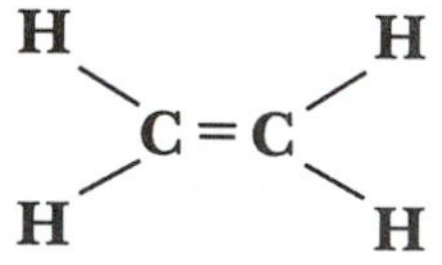

Ethylene

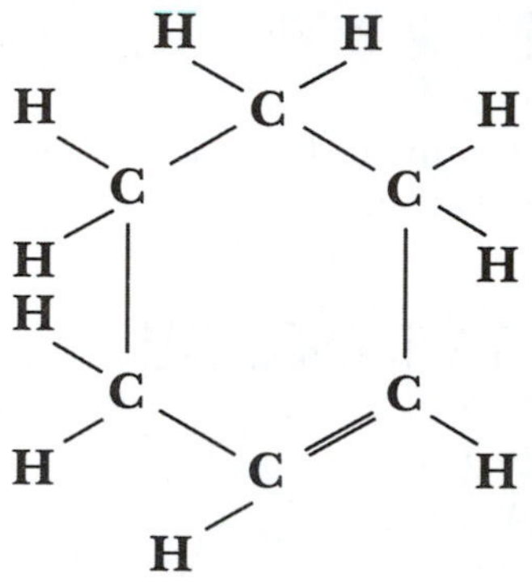

Cyclohexene

The double bond in an unsaturated compound is generally a point of greater reactivity. Among other reactions, double bonds *add* on halogens such as bromine to give saturated compounds:

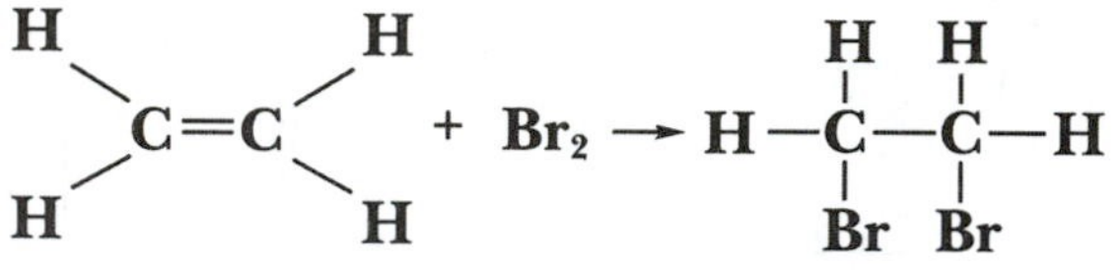

Ethylene   **Bromine**   **1, 2–Dibromoethane**

Double bonds are also attacked by oxidizing agents to give oxidized products. For example:

$$CH_3(CH_2)_2C=C(CH_2)_2CH_3 \xrightarrow{\text{Oxidizing agent}} 2CH_3(CH_2)_2C\overset{O}{-}OH$$

cis–4–Octene   **Butanoic acid**

When three alternating double bonds are present in a ring system, they are much less reactive than isolated double bonds. Such compounds are called aromatic compounds, three examples of which are

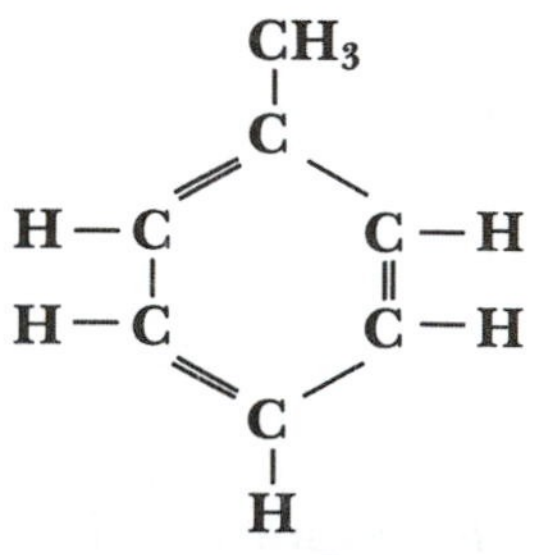

Toluene 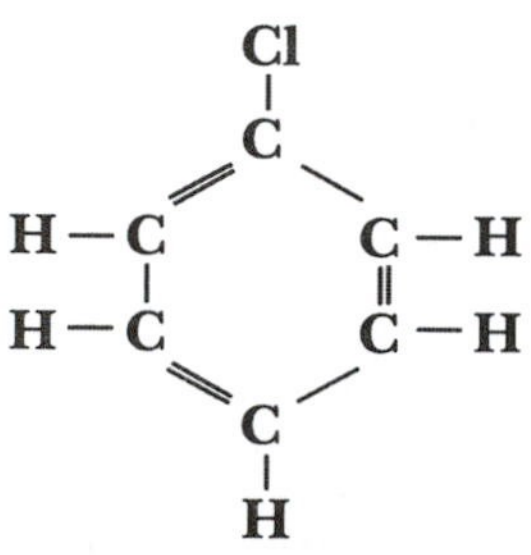   **Chlorobenzine** 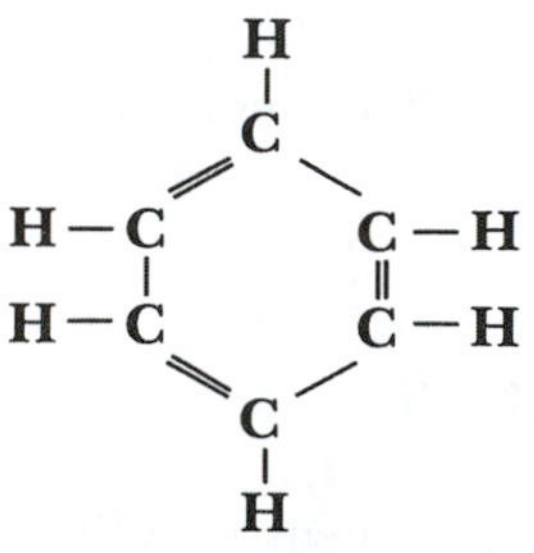   **Benzene**

Aromatic compounds have a higher carbon-to-hydrogen ratio than non-aromatic compounds. When aromatic compounds burn they often produce a sooty flame which contains unburned carbon particles.

When organic compounds containing chlorine and bromine are decomposed in a flame in the presence of copper metal, volatile copper halides are formed which give off a characteristic light. This is the basis of the Beilstein test for halogenated hydrocarbons.

PROCEDURE

Numbers in parentheses refer to entry numbers on the data sheet.

Each test is to be run with hexane, cyclohexene, toluene, and chlorobenzene. (Note your observations for the following tests on the data sheet.)

I. IGNITION Burn 1 mL of each substance on separate watch glasses under a hood. Which has the cleanest flame (1)?

II. SOLUBILITY Study the solubility of each substance in the following solvents: water (2), petroleum ether (3), and concentrated sulfuric acid (4) **DANGER!! Concentrated sulfuric acid will burn you if it gets on your skin.** For solubility, put 1 mL of solvent in a *dry* test tube. Add the substance dropwise and shake. (Add up to 1 mL of the substance.) One layer indicates solubility; two layers indicate insolubility.

III. REACTIONS (Handling Bromine Solutions is Dangerous—Be Very Careful.)
  a. *Bromine* in cyclohexane. Place 1 mL of each substance in a *dry* test tube. Add bromine in cyclohexane dropwise until color persists (5). Do not add more than 20 drops.
  b. *Potassium permanganate.* Place 1 mL of each substance in a *dry* test tube. Add permanganate dropwise. Shake well after each addition (6). Add up to 20 drops of permanganate.
  c. *Instructor demonstration.* The reaction of bromine in cyclohexane with hexane should be negative—i.e., the bromine color should not disappear. The instructor will irradiate the hexane and bromine under a bright lamp. Watch to see if there is a reaction (7). Note whether a gas is evolved. This is a substitution reaction.
  d. *Beilstein test.* Put a copper wire into the substance to be tested and then put the copper wire in a gas burner flame. Note any color change in the flame (8). This is a test for the presence of a halogen other than fluorine in an organic molecule.

PRE-LAB QUESTIONS

1. What precaution should be taken when using solvents such as hexane when Bunsen burners are also used in the same experiment?

    _____

    _____

2. What is the difference in carbon-carbon bonding between saturated hydrocarbons and aromatic hydrocarbons?

    _____

    _____

3. What is a Beilstein test, and what does a positive test indicate?

    _____

    _____

4. What are the characteristics of potassium permanganate?

    _____

    _____

Experiment 26
Data Sheet

Name _____

Date _____

THE COMPARISON OF SATURATED, UNSATURATED, AND AROMATIC HYDROCARBONS

	(SATURATED) *HEXANE*	*(UNSATURATED)* *CYCLOHEXENE*	*(AROMATIC)* *TOLUENE*	*(AROMATIC)* *CHLOROBENZENE*
(1) Ignition				
Solubility (2) Water (3) Petroleum ether (4) Sulfuric acid				
(5) Drops of bromine solution to achieve persistent color				
(6) Reaction with potassium permanganate				
(7) Reaction with bromine in bright light				
(8) Beilstein test results				

POST-LAB QUESTIONS

1. What structural feature is required in a hydrocarbon for it to react with bromine or potassium permanganate solution?

   _____

2. What is the chemical composition of the black smoke particles produced when aromatic compounds burn?

   _____

3. In determining solubility a student noticed a cloudy mixture, rather than one layer or two separate layers. Does this indicate solubility or insolubility?

   _____

27 The Preparation of Ethyl Alcohol by Fermentation*

OBJECTIVE

To examine a process in which fermentation is used to obtain one organic compound from another.

APPARATUS AND CHEMICALS

600-mL beaker
plastic sheeting to cover beaker
round-bottom flask
distilling head with small column
condenser
thermometer
receiver flasks or test tubes
boiling chips
evaporating dish
yeast

corn syrup
solution of nutrient salts for yeast
(2 g K_3PO_4, 0.2 g $Ca_3(PO_4)_2$,
0.2 g $MgSO_4$ and 10 g $NH_4C_2H_3O_2$
(dissolved in water, 900 mL))
iodine crystals
sodium hydroxide solution
cyclohexane
Bunsen burner

SAFETY CONSIDERATIONS

 Be very sure that the various components of the distillation apparatus fit together snugly. Have your instructor check your set-up before you begin the distillation. An apparatus that is not fitted or clamped properly can easily fall and/or break.

 The cyclohexane is quite flammable and must be kept away from all sources of flame. It should not be used until **all Bunsen burners have been turned off.** When using a burner, be sure that loose hair and clothing do not come close to the flame. It is easy to forget this simple rule when concentrating on measurements.

Be sure the gas is turned off completely when you are through with it since in a closed laboratory a small gas leak can create an explosive mixture.

 The NaOH solution can cause severe eye injury. Wear your safety goggles at all times. If the solution gets into your eyes, flush them immediately with cold water for several minutes and call your instructor.

 Iodine crystals will cause staining of almost anything which they contact (including skin). Use care in their use. Any NaOH solution spilled should be washed up immediately with water, followed by soapy water.

* Two laboratory periods are required for this experiment because of the period time needed for fermentation. Preparation for this experiment can be done, along with a shorter experiment, during the first period.

FACTS TO KNOW

Many chemical compounds are prepared by the action of suitable microorganisms (bacteria) on appropriate starting materials. One of the oldest of such processes is the preparation of ethyl alcohol (ethanol) from carbohydrates or sugars. Here the reaction is:

$$C_6H_{12}O_6 \xrightarrow[\text{H}_2\text{O}]{\text{Yeast}} 2\ C_2H_5OH + 2\ CO_2$$

Glucose   Ethanol
(a sugar)

As the alcohol concentration builds up, the yeasts suffer inhibition and ultimately death.

The maximum concentration of alcohol in water that most yeast can tolerate is about 12%, which is why wine, a product prepared by natural fermentation, has an alcoholic content of about 12%. The yeast literally kills itself by converting the grape sugar into alcohol.

The product of this fermentation is a mixture of yeast, water, and alcohol, from which the alcohol can be separated by distillation.

The distillation is to be carried out using the set-up illustrated in Figure 27-1.

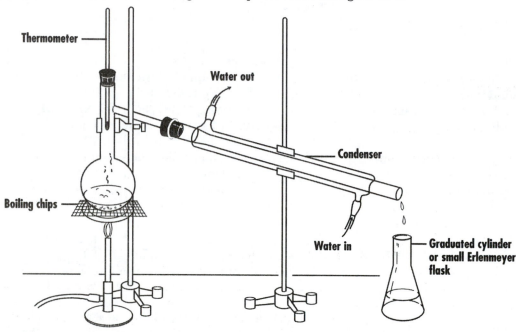

Figure 27-1. Set-up for separating ethyl alcohol and water by distillation.

A Claisen flask may be used to replace the round-bottom flask and distilling head; one is shown in Figure 27-2.

Pure ethanol boils at 78.5°C, much lower than water. When a mixture of ethanol and water is warmed, the vapor contains a higher concentration of ethanol than does the liquid. If the vapor is condensed, one can obtain a liquid with a rather high alcohol content.

Ethyl alcohol undergoes many reactions. Usually the fermentation mixture will contain some acetaldehyde and some acetic acid, both of which are formed by the oxidation of the ethanol:

$$C_2H_5OH \xrightarrow{\text{oxygen}} CH_3CHO$$

Ethanol   Acetaldehyde
(ethanal)

$$CH_3CHO \xrightarrow{\text{oxygen}} CH_3COOH$$

Acetaldehyde   Acetic acid
(ethanal)   (ethanoic acid)

In the presence of a base, iodine oxidizes ethanol to iodoform:

$$C_2H_5OH + 4\,I_2 + 6\,OH^- \rightarrow CHI_3 + HCOO^- + 5\,I^- + 5\,H_2O$$

Iodine   Iodoform   Formate   Iodide
                     (yellow)   ion

Iodoform is a yellow solid which has been used as an antiseptic.

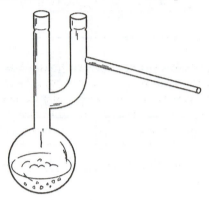

Figure 27-2. A Claisen flask.

PROCEDURE

Numbers in parentheses refer to entry numbers on the data sheet.

I. FIRST WEEK In a 600-mL beaker place 200 mL water, 30 mL corn syrup, about one eighth of a cake of yeast, and 12 mL of the nutrient salt solution. Stir the mixture thoroughly to disperse the yeast and obtain a homogeneous mixture. Place a sheeting of thin plastic over the top of your beaker and put it in your desk until the next week's laboratory.

II. SECOND WEEK Take your fermentation mixture and decant the supernatant liquid from the residue at the bottom of the flask. Set up a distillation apparatus as shown in Figure 27-1. Add your fermentation liquid to the distillation flask and add several boiling chips. At first, heat the mixture *very* gently, bringing it up to a *mere simmer!* Record the temperature at which liquid begins to distill over (1). Distill over about 15 mL of the product at as low a temperature as you can. If you heat it too strongly it will *bump* and some of the liquid will be thrown over into your receiving vessel. When you have obtained about 15 mL of distillate, remove your receiving vessel, shut off the flame, and stop the distillation. Take the 15 mL and divide it into four nearly equal portions (A, B, C, and D) for Tests 1, 2, and 3.

Test 1: (Take portion A.) Note the odor of the distillate and record your comments on your data sheet (2). After you have done this, pour *this* portion of your distillate into an evaporating dish and set a match to it. Record what happens (3).

Test 2: (Take portions B and C). In this test we will examine the ability of the distillate to dissolve in two common solvents. Two liquids are soluble if they dissolve completely in all proportions.

    To portion B add 1 mL of distilled water, swirl the mixture, and note if the two compounds are miscible. Record this on your data sheet (4).

    To portion C add 1 mL of cyclohexane, swirl the mixture, and note if the two compounds are miscible. Record this on your data sheet (5).

Test 3: To portion D add some iodine crystals and shake gently until a solution (dark brown color) is produced. Then add, *dropwise*, sodium hydroxide solution. Describe in detail what you observe (6).

PRE-LAB QUESTIONS

1. What primary hazard is associated with cyclohexane?

   _____

   _____

   _____

2. If sodium hydroxide solution comes into contact with your eyes, what should be your immediate response?

   _____

   _____

   _____

3. Since the bacteria and nutrients convert the glucose to alcohol, why couldn't you just keep increasing the amount of glucose to get higher and higher concentrations of alcohol?

   _____

   _____

   _____

4. The fermentation step is sometimes done using a soft-drink bottle as a container. If the mouth of an uninflated balloon is fitted over the opening of the bottle, is will be found to be partially inflated at the next lab period. Why does this happen?

   _____

   _____

   _____

Experiment 27
Data Sheet

Name _____

Date _____

THE PREPARATION OF ETHYL ALCOHOL BY FERMENTATION

(1) Temperature at which liquid begins to distill over from supernatant fermentation liquid.

_____°C

(2) Comments on odor of distillate

_____

_____

_____

(3) Comments on test for flammability of distillate

_____

_____

_____

Solubility of compounds in distillate:

(4) water _____

(5) cyclohexane _____

(6) Iodoform reaction description

_____

_____

_____

POST-LAB QUESTIONS

1. What property enables one to use distillation for the separation of ethyl alcohol from the water mixture?

    _____

    _____

    _____

2. Why was it necessary to "gently" distill the solution?

    _____

    _____

    _____

3. Based on your results in adding ethyl alcohol to the cyclohexane, predict the results on adding the alcohol to gasoline.

    _____

    _____

    _____

4. Enzymes operate best at a temperature of 98.7 degrees Fahrenheit (body temperature). As the enzymes convert the glucose to ethyl alcohol, heat is given off. This is especially true after the third day of fermentation. How would you design a process which would optimize the fermentation?

    _____

    _____

    _____

28 The Preparation of Organic Compounds: Aspirin and Oil of Wintergreen

OBJECTIVE

To prepare a sample of aspirin to be turned in for grading on quality and quantity.
To prepare a small sample of oil of wintergreen.

APPARATUS AND CHEMICALS

125-mL flask	filter paper
100-mL beaker	clamp
250-mL beaker	ring stand
400-mL beaker	iron ring
watch glass	burner
50-mL flask	wire gauze
10-mL graduated cylinder	concentrated sulfuric acid
dropper	salicylic acid
Büchner funnel	acetic anhydride
vacuum filtration apparatus	methanol
20-mm × 150-mm test tube	ethanol

SAFETY CONSIDERATIONS

Concentrated sulfuric acid is extremely irritating to the skin and eyes, reacts violently with water, and is destructive to clothing. Wear eye protection and handle with extreme care. Should your eyes or skin contact sulfuric acid wash with lots of water.

Acetic anhydride is also irritating to the skin and eyes. Use with adequate ventilation.

Methanol and ethanol are flammable. Methanol is toxic and breathing the vapor for short periods even in low concentration is dangerous. Swallowing small quantities of the liquid can cause blindness or death. Use with adequate ventilation.

FACTS TO KNOW

Aspirin (acetylsalicylic acid) and oil of wintergreen (methyl salicylate) are commonly used organic compounds. Oil of wintergreen is used in rubbing liniments for sore muscles and as a flavoring agent. Aspirin is used as an analgesic (pain reliever) and as an antipyretic (fever reducer). Aspirin tablets (5 grain) are usually compounded of about 0.32 g of acetylsalicylic acid pressed together with a small amount of starch which binds the tablet together.

Aspirin is usually made by reacting salicylic acid with acetic anhydride. Acetylating agents other than acetic anhydride can be used; however, acetic anhydride is used because it is inexpensive and is not too difficult to handle safely in the laboratory. Aspirin is a white solid and is almost insoluble in water. Only 0.25 g will dissolve in 100 mL of water. Thus, aspirin can be separated from the reaction mixture by crystallization and filtration. Sulfuric acid is used in this experiment as a **catalyst,** a substance that increases the rate of a chemical reaction without being permanently changed itself.

Salicylic acid   Acetic anhydride   Aspirin   Acetic acid

Oil of wintergreen is made by reacting salicylic acid with methanol in the presence of sulfuric acid catalyst.

Oil of wintergreen has a pleasant odor, which is characteristic of organic esters.

Salicylic acid   Methanol   Oil of Wintergreen

PROCEDURE

I. PREPARATION OF ASPIRIN Set up a water bath as shown in Figure 28-1. Place 6 g (weigh accurately to 0.01 g) of salicylic acid in a 125-mL conical flask. Add 8 mL of acetic anhydride to the salicylic acid in the conical flask. *CAUTION!!* While swirling the flask, add 10 drops of concentrated sulfuric acid to the mixture. Heat the flask in a beaker of boiling water (Fig. 28-1) for 15 minutes. If the solid does not dissolve, heat 10 more minutes. You may need to stir in the final stages. You will not get a good yield of aspirin unless all of the salicylic acid dissolves at this point. Remove the flask from the water bath and add 25 mL of ice water to the flask. Set the flask in a beaker of ice until crystallization appears to be complete (Fig. 28-2). Separate the crystals from the liquid by vacuum filtration (Fig. 28-3).

Recrystallize the aspirin by dissolving the crystals in 20 mL of *ethanol** in a 100-mL beaker. Warm on a water bath (Fig. 28-1) if necessary to effect solution. Then pour 50 mL of warm water into the solution. Cover the beaker with a watch glass and set aside to cool.

The beaker may be set in a beaker of ice to speed cooling (Fig. 28-2).

* Be certain you use ethanol here. Two alcohols will be in the lab. Only ethanol works well at this point.

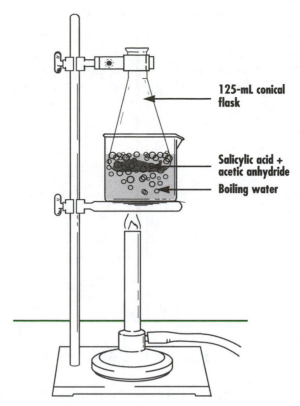

Figure 28-1. Apparatus for heating the reaction mixture.

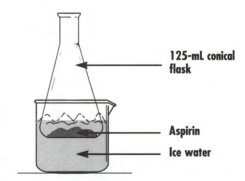

Figure 28-2. Set-up for cooling the reaction mixture.

At this point you may prepare the oil of wintergreen sample, Part II.

Filter the aspirin crystals, using the vacuum filtration apparatus (Fig. 28-3). Dry them by spreading them on a piece of filter paper (or paper towel) and patting them with another piece of filter paper. Stir and chop with a spatula. Weigh the aspirin in a preweighed 50-mL flask. A 100% yield of aspirin would produce 7.82 g of aspirin. Calculate your percent yield using the following formula:

$$\% \text{ yield} = \frac{\text{weight aspirin}}{7.82 \text{ g}} \times 100$$

Place the following information on the bottle label and turn the product in to your instructor:

Name of student
Name of product
Yield in grams
Percent yield

No report sheet is required for this experiment.

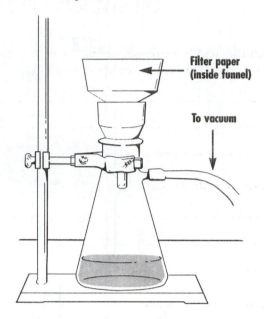

Figure 28-3. Apparatus for filtering the crystals.

II. PREPARATION OF OIL OF WINTERGREEN   Carry out the preparation of oil of wintergreen in the hood. Place about 0.2 g of salicylic acid in a medium-sized test tube (20 mm × 150 mm). Add 2 mL of methanol and swirl until the solid dissolves. Cautiously add 5 drops of concentrated sulfuric acid to the mixture and place the test tube in a hot water bath at 60°C to 70°C for 15 minutes. Then remove the test tube from the hot water bath and allow the test tube to cool to room temperature. Cautiously smell the contents of the test tube by moving your cupped hand over the test tube toward your nose (Fig. 28-4). Show the mixture to your instructor.

Figure 28-4. Exercise great care in noting the odor of a substance, using your hand to waft its vapor gently toward your face. Whenever possible, avoid breathing fumes of any kind.

PRE-LAB QUESTIONS

1. Concentrated sulfuric acid is extremely irritating to the skin and eyes. If your skin or eyes contacted this acid, what first aid procedure would you follow?

    _____

    _____

    _____

2. What is the purpose of sulfuric acid in this reaction?

    _____

    _____

    _____

3. Oil of wintergreen is found to be present in many sports cremes. Speculate as to whether it may have some other affect other than providing a pleasant odor.

    _____

    _____

    _____

4. What alcohol should be used to recrystallize aspirin?

    _____

POST-LAB QUESTIONS

1. Why did you use a water bath instead of a Bunsen burner in the recrystallization of the aspirin?

   _____

   _____

   _____

2. If your aspirin after recrystallization was not dry, how would this affect your yield? (High or low, why?)

   _____

   _____

   _____

3. Based on the yield that you got in your experiment, how much salicylic acid would you need to produce 10.0 g of aspirin? (Show calculations)

4. The instructions called for you to "recrystallize" the aspirin from ethanol (ethyl alcohol). What does recrystallization accomplish?

   _____

   _____

   _____

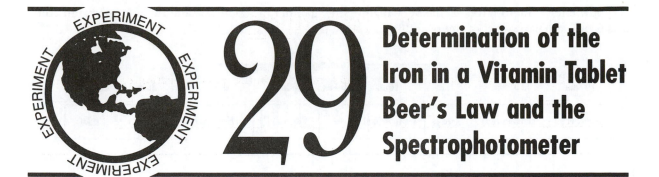

Determination of the Iron in a Vitamin Tablet Beer's Law and the Spectrophotometer

OBJECTIVE

To demonstrate the use of a spectrophotometer to do quantitative analysis of substances that absorb electromagnetic radiation. To learn the principles of Beer's Law.

APPARATUS AND CHEMICALS

Spectrophotometer and sample tubes (cells)
1000 mL volumetric flask
250 mL volumetric flask
100 mL volumetric flask
10 mL pipette
50 mL pipette
7 (25 mL capacity) small c
small funnel

Stock solutions:
2 M H_2SO_4
1.4 M hydroxylamine
0.0056 M orthophenanthroline
1.2 M $NaC_2H_3O_2$
2.80×10^{-4} M $Fe(NH_4)_2(SO4)_2$
25 mL graduated cylinder

SAFETY CONSIDERATIONS

Sulfuric acid can cause serious eye damage. Wear safety glasses at all times.

Pipettes and volumetric flasks are particularly fragile. Handle them carefully and be very careful with broken glass

EXPERIMENT

SPECTROPHOTOMETRIC ANALYSIS OF IRON IN A VITAMIN TABLET

Some of the most powerful tools the analytical chemist has at his or her command are a class of instruments known as spectrophotometers. These instruments can be used for qualitative analysis (to identify an unknown substance or the components of a mixture), or for quantitative analysis (to measure amounts of materials present in a mixture). The instruments use electromagnetic radiation ("light") as the agent of analysis, hence the name *spectro* (from spectrum), *photo* (light photon), *meter* (measure).

Electromagnetic radiation, or light as it is commonly called, interacts with matter in a variety of ways, and as a result is absorbed by matter. The pattern of absorbance speaks volumes about the nature of the matter being investigated. Figure 29-1 shows the familiar electromagnetic spectrum, from the very low energy radio waves to the very high energy gamma rays. The so-called visible spectrum comprises only a narrow part of the overall spectrum.

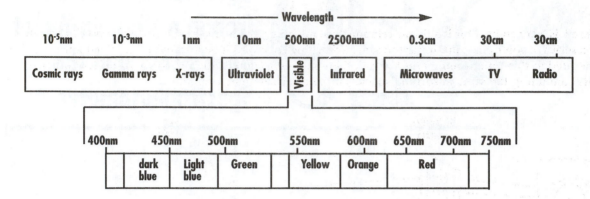

Figure 29-1. Electromagnetic spectrum.

How light interacts with matter depends on the energy of the light in question. Radiowaves, for example, cause small changes in the magnetic fields of the nuclei of some atoms which allows the determination of the environment that the nucleus is in. This helps analysts to discover structural features of molecules. Microwaves cause molecules to rotate very rapidly, and from the rotational patterns, analysts can determine bond angles and bond lengths in molecules. Infra-red light causes the bonds in molecules to vibrate, much like springs, and thus the bonding features of a molecule may be studied. Finally, visible and ultra-violet radiation cause electrons in molecules to be excited into higher energy levels. These areas of the spectrum are used most often for quantitative analysis of amounts of material present, as is the case in this experiment.

When the light being absorbed falls within the range of visible light, the material will appear colored. White light contains all visible wavelengths (this is termed polychromatic light). When one or more wavelengths of visible light is absorbed, the remaining light is colored, and the color observed is said to be the compliment of the absorbed color. Blue objects exhibit that color because they absorb light in the red region of the spectrum, and reflect the other regions to be seen by the observer.

When clear solutions absorb light they also take on color. **IT IS IMPORTANT TO UNDERSTAND THE DIFFERENCE BETWEEN THE TERMS CLEAR AND COLORLESS!** *Clear* means that one can see through it (even though it may have color). *Colorless* means there is no color, but does not necessarily mean it can be seen through. Milk is colorless, but is not clear. Water is clear and colorless. Lime Koolaid is clear, but not colorless. If a solution is not clear, then light cannot effectively pass through it. These solutions are said to be turbid or even opaque. In order for solutions to be effectively analyzed by spectrophotometry, they must be clear, although they are frequently colored.

The absorbance of light by a solution is governed by a natural law known as Beer's Law. Beer's Law states that the absorbance of light is proportional to the concentration of the dissolved species absorbing the light and to the distance that the light travels through the solution. The term absorbance is related to the quantity of light that enters the solution compared to the quantity exiting the solution. Mathematically, Beer's Law is given by equation 29-1, where C equals the concentration of the absorbing species, b is the path length of light through the sample, and a is the proportionality constant.

$$Abs = a\,b\,C \qquad\qquad\qquad 29\text{-}1$$

The proportionality constant is a characteristic of the particular species that is absorbing the light, as well as the wavelength of light being absorbed. In normal analysis, one compares solutions which all contain the same absorbing species (at a constant wavelength), thus a does not change. Likewise, the solutions are compared in containers of identical size, so that the path length b does not change. This makes the absorbencies of the solutions directly proportional to (only) their concentrations.

$$\frac{Abs_1}{Abs_2} = \frac{a\,b\,c_1}{a\,b\,C_2} \quad \text{or} \quad \frac{Abs_1}{Abs_2} = \frac{C_1}{C_2} \quad \text{(since a \& b are constant)}$$

Another way to express the relationship between like solutions in the same sized container (a & b do not change) is

$$ABS = k \, c$$

where k is a constant. This is a linear relationship, meaning that a graph of absorbance vs concentration for a series of solutions containing the same absorbing species will be a straight line. Such a line will be generated in this experiment. This line can then be used to determine the concentration of other solutions containing the same absorbing species, if their absorbencies are known.

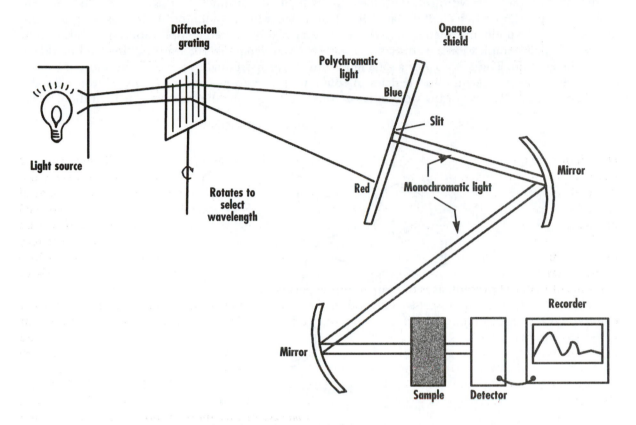

Figure 29-2

A schematic diagram of a spectrophotometer is shown in Figure 29-2. It consists of (1) a light source (a tungsten bulb for visible light); (2) a monochromator, which separates the polychromatic light into light of a single wavelength (monochromatic light); (3) a sample holder through which the light passes for analysis; and (4) a detector which measures the intensity of the light beam at the end of its path. Of course there are many additional parts such as photomultipliers, filters, mirrors, etc., to direct the light beam and optimize the performance of the instrument, but which do not alter the method of operation. Of the four major parts mentioned above, only the monochromator needs further explanation. Consider visible light passed through a prism and cast onto a screen. It produces the familiar continuous spectrum we recognize as a rainbow, which contains all wavelengths of visible light. Suppose the screen has a very narrow slit in one place. The light that gets through the slit consists of a very narrow band of wavelengths, or even of a single wavelength if the slit is narrow enough. This is monochromatic light. If the prism is slowly rotated on its axis, the visible spectrum moves across the screen, and different wavelengths of light get through the slit. By controlling the angle of the prism, one can select the wavelength of light that gets through the slit and to the sample. Modern instruments do not use prisms to disperse the light into the spectrum, but instead us *diffraction gratings* to accomplish the same thing. A diffraction grating is a piece of translucent material with thousands of parallel lines scored into its surface (up to 30,000 per inch), and which produces the same spectra that prisms produce.

SPECTROPHOTOMETRIC DETERMINATION OF IRON

Many vitamin tablets contain iron. The iron is needed by the body for the production of hemoglobin, which carries oxygen in the bloodstream (see the discussion for Experiment 18). Iron provided by a tablet must be in water soluble form in order for the body to use it, thus the iron can be analyzed in an aqueous solution by dissolving the tablet. Some vitamin tablets contain iron inits elemental form—that is, as the powdered metal. This iron must be dissolved, which stomach acids do quite nicely! Iron does not absorb light very strongly by itself insolution, so a reagent is added to the solution that produces a complex ion with iron that does absorb light very strongly. In this way, very low concentrations of iron can be detected and analyzed. Iron reacts with a compound known as orthophenanthrolene (or 1,10-phenanthrolene) to produce the very highly colored complex ion which is shown in Figure29-3. This complex ion absorbs light strongly at a wavelength of 508 nm, and thus looks reddishorange in solution. The absorbance at 508 nm is directly proportional to the concentration of iron in the solution. By comparing absorbencies of solutions with known concentrations of iron with the absorbance of the solution of the vitamin tablet, one can determine the amount of iron in the tablet.

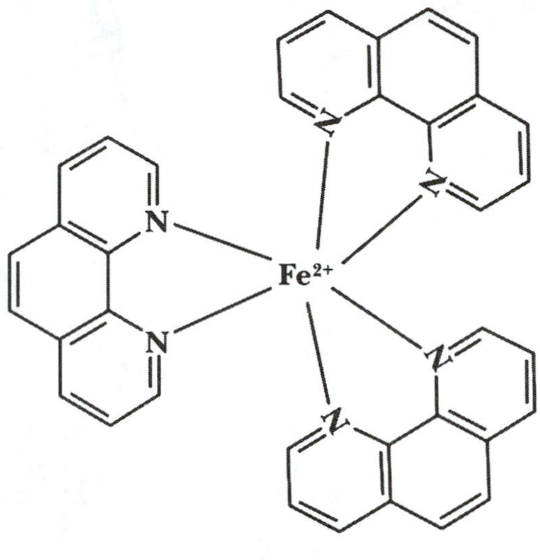

Figure 29-3

THIS EXPERIMENT IS DESIGNED FOR GROUPS OF 3 OR 4 WORKING TOGETHER.

In this experiment, you are to prepare several solutions containing exact amounts of either iron (Fe^{2+}) or unknown material (from a vitamin tablet). In each case, the solutions you are to prepare are made by transferring measured amounts of several "stock" solutions into a volumetric flask and carefully diluting the resulting solution to the mark on the neck of the flask. In all cases, the quantity of the "stock" iron solution to be used must be measured very carefully—with a pipette. The other stock solutions are all being used in excess (the iron is always the limiting reagent), thus the measurements of these quantities are not as critical and may be done with a graduated cylinder. The dilution in the volumetric flask must be done very carefully!

1. Weigh, on the analytical balance, a whole vitamin tablet which contains iron (1). Grind the tablet into a fine powder in a mortar and pestle. Using a small pre-weighed beaker (2), weigh out approximately 0.2 grams of the pulverized tablet on the analytical balance (3,4). Add 10 mL of 2 M sulfuric acid to the beaker and warm the mixture gently with your burner. Check the mixture for black iron fillings. These should all dissolve. transfer this mixture to a 1000 mL volumetric flask, taking care to rinse the beaker several times, adding the rinsings to the volumetric flask. Fill the flask about 1/2 full with deionized water, add 10 of 2.0 m sulfuric acid, and dilute to the mark. Mix this solution well. Because vitamin tablets frequently contain starch or other water insoluble binders, this solution may appear slightly cloudy or translucent. Since spectrophotometric analysis depends on the solution being transparent, it is necessary to filter the solution. The filtration must be done into a clean *and dry* beaker, so as to not change the concentration of the solution. only about 200 mL of this solution needs to be filtered and saved.

2. Pipette 50 mL of the solution of the vitamin tablet into a 100 mL volumetric flask, add 1.5 mL of hydroxylamine solution, 10 mL of orthophenanthroline solution, 10 mL of sodium acetate solution and 4 mL of 2 M H_2SO_4. Dilute to the mark and mix well. Save 25 mL.

3. Prepare four (4) standard solutions of iron-phenanthroline complex for use in making a calibration curve. Pipette 10 mL of the stock iron solution provided by your instructor into a 250 mL volumetric flask. Add 1.5 mL of the stock hydroxylamine solution, 10 mL of the stock sodium acetate solution and 4 mL of 2 M H_2SO_4. Dilute to the mark and mix the solution well. Prepare the other 3 standard solutions in exactly the same way, using 20 mL, 30 mL, and 40 mL pipetted aliquots of the stock iron solution, respectively. All amounts of the other reagents are the same. Save about 25 mL of each standard solution in a clean, dry container.

4. Prepare a blank solution exactly the same way that the 4 standard solutions were prepared, *omitting only the stock iron solution*. As before, dilute carefully to the mark on the neck of the flask. Save 25 mL of this solution.

5. Allow the 4 standard solutions and the solution of the tablet material to stand for at least 5 minutes, then measure the absorbance of the 4 standard solutions and of the tablet solution at 508 nm, using the blank solution as your reference.

CALCULATIONS

The concentration of the iron/phenanthroline complex will be equal to the concentration of the iron itself, since the phenanthroline is added in excess and the iron is essentially 100% complexed in these solutions.

$$\text{Stock Iron solution: } (1000 * \text{Mass FAS}) \times \frac{5.85 \text{ g Fe}}{392.15 \text{ g FAS}} = C_1 \text{ (mg Fe/L)*}$$

(Your instructor may provide this solution and give you the value for C_1.)

* FAS = ferrous ammonium sulfate; formula weight = 392.15 g/mol
* The concentration unit mg/L is also known as parts per million (ppm).
* Concentrations of standard iron solutions are calculated as follows, starting with C_1:

$$C_1 \times \text{Vol pipetted} = C_2 \times 250 \text{ mL}$$
$$C_1 \times \text{Vol pipetted} = C_3 \times 250 \text{ mL}$$
$$C_1 \times \text{Vol pipetted} = C_4 \times 250 \text{ mL}$$

Prepare a graph of absorbance (y axis) vs concentration (x axis) for the 4 standard solutions, and draw the best straight line through the points. From the absorbance of the tablet material solution, read the concentration of iron in this solution (C_{Fe}) from the graph. This has units of mg Fe/L.

The total milligrams of iron in the tablet is calculated as follows, based on the dilutions made and the original mass of the tablet.

$$C_{Fe} \times 1.00 \text{ L} \times \frac{100 \text{ mL}}{50 \text{ mL}} \times \frac{\text{mass of original tablet}}{\text{Mass of powdered sample}} = \text{mg Fe in tablet}$$

There are many different models of spectrophotometers that may be encountered. One of the most common models used in introductory laboratory courses is the Bausch & Lomb Spectronic 20. (Figure 29-4) Because it measures light in the visible range (colored light), it is frequently called a colorimeter. Directions for the use of the Spectronic 20 are as follows.

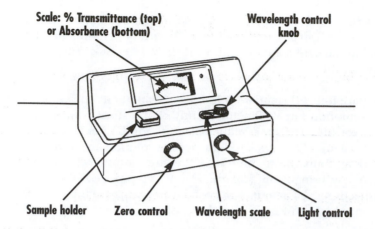

Figure 29-4. Spectronic 20 spectrophotometer.

1. Turn on the instrument and allow it to warm up for at least 30 minutes. The instrument should be turned on at the beginning of the laboratory period.

2. Set the spectrophotometer wavelength control to 508 nm.

3. Using the zero control knob, adjust the indicator needle on the left side of the scale to zero. The sample holder is empty and the light box is closed.

4. Obtain a sample tube from your lab instructor. Fill the sample tube about two-thirds full with deionized water, wipe the outside of the tube carefully with tissue paper, and insert it into the sample holder.

5. Adjust the indicator needle on the right side of the scale until the meter reads 100 (% transmittance using the light control knob). The absorbance should read zero.

6. Fill the sample tube about two-thirds full with the most dilute solution after rinsing the tube with 1 mL to 2 mL of the solution. Place the sample tube in the sample holder and read the absorbance. Record this value (1a).

7. Repeat step 6 with each solution, continuing to measure solutions in order of increasing concentration.

PRE-LAB QUESTIONS

1. What is the function of the monochromator in a spectrophotometer?

   _____

   _____

   _____

2. On what three things does the absorbance of a solution depend?

   _____

   _____

   _____

3. How is the experiment structured so that the absorbencies of the solutions depend only on one of the three things mentioned in question 2, and which is this?

   _____

   _____

   _____

4. Why must the solution containing the powdered vitamin tablet be filtered prior to use, and what would be the effect on the final results if it were not filtered?

   _____

   _____

   _____

Experiment 29
Data Sheet

Name _____

Date _____

SPECTROPHOTOMETRIC DETERMINATION OF IRON IN A VITAMIN TABLET

DATA SHEET

1. Weight of vitamin tablet   _____ g

2. Weight of empty beaker   _____ g

3. Weight of beaker plus pulverized tablet   _____ g

4. Weight of powder (subtract 2 from 3)   _____ g

TABLE OF ABSORBENCIES

Solution	Concentration (mg/L)	Absorbance
Iron standard #1	_____	_____
Iron standard #2	_____	_____
Iron standard #3	_____	_____
Iron standard #4	_____	_____
Vitamin tablet solution	_____ (from graph)	_____

Total milligrams of iron in vitamin tablet   _____ mg

POST-LAB QUESTIONS

1. What would be the effect on the final analysis if an insufficient quantity of the orthophenan-
    throline solution had been added to the standard solutions, and why?

    _____

    _____

    _____

2. What would be the effect on the final analysis if the spectrophotometer cell had fingerprints on it
    when the absorbencies of the standards were measured, but not when the absorbance of the
    unknown was measured, and why?

    _____

    _____

    _____

3. Absorbencies greater than about 1.5 cannot be measured as accurately as those below 1.0, since
    at an absorbance of 1.5, 97% of the light entering the solution has been absorbed and it is difficult
    to measure the small amount of light exiting the solution. What simple steps could have been
    done if the absorbance of the unknown exceeded 1.5? (Hint: think of the procedure used in the
    experiment.)

    _____

    _____

    _____

Identification of Certain Types of Organic Compounds

OBJECTIVE

To identify functional groups of organic compounds by performing chemical tests.
To identify an organic compound.

APPARATUS AND CHEMICALS

small test tubes (12)
50-mL beakers (3)
stirring rod
medicine dropper
litmus paper
test tube rack
test tube holder
burner and hose
cooking fat
pentene

ethanol
1-butanol
t-butyl alcohol
sodium
Fehling's solution A and Fehling's solution B
acetaldehyde
acetone
acetic acid, glacial
dodecylamine
5% bromine in cyclohexane

SAFETY CONSIDERATIONS

    Wear eye protection to protect against splattering of chemicals and possible shattering of glassware.

    Sodium is very reactive with water. It produces hydrogen which explodes in the presence of a spark or flame. Large amounts of sodium with water can be explosive, especially in a confined space.

    Bromine is a corrosive poison and can cause deep burns on human skin.

Dodecylamine is an irritant and should not be breathed nor allowed on the skin. Vapors of glacial acetic acid are irritating to the eyes, nose, and skin.

    Cyclohexane, pentene, acetone, and alcohols are flammable and should not be placed close to an open flame.

    Protect your hair, clothing, desk, and books from the open flame of a burner.

FACTS TO KNOW

Some parts of the molecular structure of organic compounds can be identified by simple chemical tests. Very often, the total structure of organic compounds is not determined; only the presence of reactive groups (functional groups) is established. Once the reactive groups are identified, physical and chemical properties are used to identify a specific organic compound.

Small samples of an unknown substance can be subjected to tests with a number of reagents in order to detect the presence or absence of reactive groups. Some important functional groups examined in this experiment are

| Unsaturated group | Aldehyde or ketone carbonyl group | Hydroxyl group | Carboxyl group |

R indicates the rest of the organic molecule.

Unsaturated organic compounds contain one or more carbon-carbon double bonds (C=C). A double bond can be detected by using either bromine in cyclohexane or aqueous potassium permanganate. Bromine will react with unsaturated compounds, and a positive test is easily detected by the loss of the red-brown bromine color.

Ethylene   (Red-brown)   (Colorless)

There are many compounds that contain the **carbonyl group,** such as aldehydes, ketones, and carbohydrates (sugars). One of the many chemical tests for the carbonyl group is the Fehling's test which is a test for **aldehydes.** In the presence of an aldehyde group, the Fehling's reagent turns from blue to green and gradually to a reddish precipitate of cuprous oxide (Cu_2O). Simple ketones such as acetone do not show a positive test with Fehling's solutions.

Acetone

Organic acids have the carboxyl group (−COOH) and can be detected by placing the organic acid in water and testing with litmus paper. Blue litmus paper turns red in the presence of an acid.

$$R-\overset{\overset{\displaystyle O}{\|}}{C}-OH + H_2O \rightleftharpoons R-\overset{\overset{\displaystyle O}{\|}}{C}-O^- + H_3O^+$$

Organic bases have the amine group (−NH₂) and can be detected in water by using red litmus, which turns blue.

$$R-NH_2 + H_2O \rightleftharpoons RNH_3^+ + OH^-$$

Organic compounds containing the hydroxyl group (−OH) are commonly called alcohols. Alcohols are classified as primary, secondary, or tertiary, depending on whether the hydroxyl group is attached to a carbon that is bound to one, two, or three other carbon atoms, respectively.

Primary alcohol

Secondary alcohol

Tertiary alcohol

Sodium will evolve bubbles of hydrogen gas in the presence of alcohols and can be used as a test for the hydroxyl group **only if the substance does not affect litmus.** *Wear safety glasses when you perform this test!*

$$2R-O-H + 2Na \longrightarrow 2R-\overset{\displaystyle H}{\underset{\displaystyle H}{C}}-O^- + 2Na^+ + H_2 \text{ (gas)}$$

PROCEDURE

Numbers in parentheses refer to entry numbers on the data sheet.

I. TEST FOR UNSATURATION   Obtain 5 mL of pentene and 5 mL of melted cooking fat. Perform the following test for unsaturation on these compounds, using a small test tube.

   To a portion of each sample, add 5 drops of 5% bromine in cyclohexane and shake the mixtures. Note the disappearance or gross fading of color, if any (1).

   For what groups are these reagents a test (2)?

II. TEST FOR ALDEHYDE GROUP   Place 3 mL of Fehling's solution A and 3 mL of Fehling's solution B in a test tube and add 5 drops of acetaldehyde. Shake the test tube to mix the ingredients. Place the test tube in boiling water for 2 or 3 minutes and observe the specific color changes (3). Repeat the test with a ketone, such as acetone (4). Can aldehydes be differentiated from ketones by this test (5)?

III. TEST FOR ACIDIC AND BASIC GROUPS   Obtain small samples of acetic acid and dodecylamine. Test each of these for solubility in water by adding a few crystals or drops of the organic substances to 3 mL of water. Record your results (6). Substances soluble in water may be tested with litmus paper. Test all substances with both kinds of litmus. What are the color changes in the litmus (7)? Acid or base (8)?

IV. TEST FOR THE HYDROXYL GROUP WEAR SAFETY GLASSES   Obtain 5-mL samples of the representative alcohols—ethanol, n-butyl alcohol, and t-butyl alcohol. Note their characteristics, such as color, odor, and solubility in water (9). Add a tiny piece of sodium to each alcohol (free from water) and note the results (10). What are the products of this reaction (11)? **Note: Organic acids and water react violently with sodium. Alcohols do not affect litmus, while acids turn litmus red. Keep this in mind when trying to determine your unknown.**

DO NOT ADD SODIUM TO ANY SOLUTION OR SUBSTANCE THAT TESTS ACIDIC, OR TO WATER!

V. TESTS FOR THE FUNCTIONAL GROUPS IN AN UNKNOWN   Obtain an unknown organic substance from the instructor and, by means of the tests described, determine whether the substance contains unsaturated bonds, acidic or basic groups, alcoholic hydroxyl groups, or an aldehyde. Keep in mind that several of these groups may be present in the molecule of a single substance. Do not assume, therefore, that a group is absent unless the test for it is negative. Indicate your observations of the tests performed and list the functional groups you identified in your unknown.

PRE-LAB QUESTIONS

1. What is the primary hazard associated with using sodium metal?

   _____

   _____

   _____

2. What type of test would you run to determine whether or not fat was a saturated or unsaturated fat?

   _____

   _____

   _____

3. In view of the fact that sodium reacts very aggressively with water and organic acids, what test should you perform on your unknown before adding metallic sodium?

   _____

   _____

   _____

4. Natural oils (ex. olive oil, coconut oil, corn oil) are often referred to as "polyunsaturated" and are thought to be healthier for consumption than animal fats, which are "saturated." What does "polyunsaturated" mean, and why might this be a healthier product?

   _____

   _____

   _____

Experiment 30   Name _____
Data Sheet   Date _____

IDENTIFICATION OF CERTAIN TYPES OF ORGANIC COMPOUNDS

I. UNSATURATION

(1) Addition of 5% bromine in cyclohexane to:

pentene _____

melted fat _____

(2) _____

II. ALDEHYDE GROUP

Observations:

(3) Addition of Fehling's solution to acetaldehyde _____

(4) Addition of Fehling's solution to acetone _____

(5) _____

III. ACID AND BASE GROUPS

COMPOUND	SOLUBILITY IN WATER (6)	LITMUS COLOR (7)	ACID OR BASE (8)
Acetic acid			
Dodecylamine			

IV. ALCOHOLS

COMPOUND	SOLUBILITY IN WATER (9)	SODIUM TEST (10)
Ethyl alcohol		
n-Butyl alcohol		
t-Butyl alcohol		

(11) Products:

V. UNKNOWN

A. Observations:

    1. Unsaturation _____

    2. Aldehyde group _____

3. Acidic or basic group _____

4. Hydroxyl group _____

B. Functional groups present

_____

_____

_____

POST-LAB QUESTIONS

1. What tests would you use to distinguish between methyl amine and acetic acid?

    _____

    _____

    _____

2. Would you expect an organic acid to react with metallic sodium? Write the chemical equation.

    _____

    _____

    _____

3. An unknown produced the following tests: a neutral pH (not acidic or basic), gave a positive Fehling test and a positive metallic sodium test. Which functional groups would be consistent with these data? Justify your answer.

    _____

    _____

    _____

4. What two different conclusions could one come to if an unknown were tested with red litmus paper and the paper remained red?

    _____

    _____

    _____

31

Preparation and Properties of a Soap

OBJECTIVE

To prepare a sample of a soap and to examine its properties.

APPARATUS AND CHEMICALS

250-mL beaker
100-mL beaker
wire gauze
laboratory burner
glass stirring rod
test tubes
filter flask and Büchner funnel
filter paper
cooling oil
graduated cylinder

sodium hydroxide, 20% solution
ethanol
saturated solution of sodium chloride
calcium chloride, 5% solution
magnesium chloride, 5% solution
ferric chloride, 5% solution
kerosene
phenolphthalein indicator solution
graduated cylinder

SAFETY CONSIDERATIONS

Sodium hydroxide solution is extremely harmful to the skin and to clothing. It is especially harmful to the eyes. Wear safety glasses. Any sodium hydroxide spilled on the skin should be washed off immediately with large quantities of water.

Kerosene and ethanol are flammable. Care should be taken when using these chemicals around flames.

FACTS TO KNOW

A soap is the sodium or potassium salt of a long-chain fatty acid. Most solid soaps are sodium salts of the type to be made in this experiment. The starting materials are fats or oils, which are the glycerol esters of the fatty acids. A typical cooking oil, cottonseed oil, and a concentrated solution of sodium hydroxide are the principal reactants. Ethanol is used in this experiment to serve as a common solvent for the reactants and hence to speed up the reaction. It is not used in commercial soap-making operations.

The soap-making, or saponification, reaction which occurs here can be written as

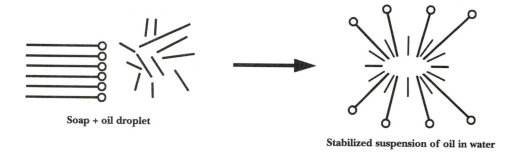

An oil (a glyceride or ester   glycerol   Sodium salts of the fatty acids
   of fatty acids and glycerol)   present in the glycerides

The sodium salts of the long-chain fatty acids are typical surface-active agents. They have a polar end which is hydrophilic (water-loving) and a long nonpolar chain which is hydrophobic (water-hating). As a consequence they can form emulsions by suspending oil in water when their nonpolar end is in the oil and their polar end group is in water (Fig. 32-1). The emulsification of kerosene in water by means of soap is just such a process.

Soap + oil droplet

Stabilized suspension of oil in water

Figure 31-1. Soap emolsifies oil.

When present, calcium, magnesium, iron, and some other metal salts can form insoluble precipitates with the long-chain fatty acids. This kind of reaction is responsible for the problems which arise when soap is used with hard water (water containing appreciable amounts of dissolved calcium, magnesium, or iron salts).

PROCEDURE

Numbers in parentheses refer to entry numbers on the data sheet.

Measure 20 g of cottonseed or other vegetable oil into a 250-mL beaker. Add 20 mL of ethanol and 25 mL of 20% sodium hydroxide solution. The mixture is stirred in the beaker. The beaker is placed on a wire gauze supported on a ring stand and heated gently. *PRECAUTION!* Heat the alcohol-oil-sodium hydroxide solution gently; keep the flame away from the top of the beaker to prevent the alcohol from catching fire as the alcohol escapes. Continue to stir as the heating is continued. Cup your hand and waft the vapors to your nose. Do not let your hair or clothing get near the flame. The heating is continued until the odor of alcohol has disappeared and a pasty mass remains in the beaker. This pasty mass is a mixture of the soap and the glycerol freed in the reaction. Turn off the burner.

Wrap the beaker with a paper towel and set the beaker (with the soap) on the lab desk to cool.

Add 100 mL of a saturated sodium chloride solution to your soap preparation and stir the mixture thoroughly with a glass stirring rod. This process is called "salting out" and is used to remove the soap from water, glycerol, and any excess sodium hydroxide present. After the mixture has been stirred and mixed completely, filter off the soap with a vacuum filtration apparatus and wash once with ice water. Get the soap as dry as possible. Cut up the soap with a spatula while air is being pulled through the Büchner funnel. Weigh your dried soap and record the weight (1).

WASHING PROPERTIES   Take a small amount of soap and try to wash your hands with it. It should lather readily if soft water is used. If soft water is not available in your area, use deionized water here. If too much oil was used, the soap will feel greasy; if too much sodium hydroxide is still present, it will also make the hands feel slick but will roughen them. Record your observations (2).

EMULSIFICATION   Put 5 to 10 drops of kerosene in a test tube containing 10 mL water and shake. An emulsion or suspension of tiny oil droplets in water will be formed. Let this stand a few minutes. What happens (3)? Prepare another test tube with the same ingredients but also add a small portion ($1/2$ g or so) of your soap before you shake it up thoroughly. Compare the relative stabilities of the two emulsions. Does the soap have any effect (4)?

HARD WATER REACTIONS   Take 1 g of your soap and warm it with 50 mL of water in a 100-mL beaker. When you have obtained a reasonably clear solution, pour about 15 mL into each of three test tubes. Test one of the three test tubes with 10 drops of 5% $CaCl_2$ solution, one with 10 drops of 5% $MgCl_2$ solution, and one with 10 drops of 5% $FeCl_3$ solution. Let these solutions stand until you have finished the other tests. Then, make your observations (5 through 7).

BASICITY   A soap with free alkali can be very damaging to skin, silk, or wool. Dissolve a small piece of your soap in 15 mL of ethanol and then add two drops of phenolphthalein. If the indicator turns red, the presence of free alkali is indicated. What happens (8)?

Turn in your soap sample. Put the following information on the label:
    Your name
    Product name
    Yield in grams

PRE-LAB QUESTIONS

1. State one safety precaution to be taken when using:

    (a) Sodium hydroxide solution _____

    (b) ethanol _____

2. Describe how to determine safely if all ethanol has evaporated from the reaction mixture.

    _____

3. What class of natural compounds serves as the starting material for the preparation of soap when reacted with NaOH?

    _____

    _____

4. How is the soap recovered as a solid product from the solution containing the other products of the reaction?

    _____

    _____

Experiment 31
Data Sheet

Name _____

Date _____

PREPARATION AND PROPERTIES OF A SOAP

I. PREPARATION

(1) Yield of dry soap   _____ g

II. PROPERTIES (No pronouns, please)

(2) Washing properties _____

_____

_____

Emulsification of oil:

(3) Behavior of oil and water mixture _____

_____

_____

(4) Behavior of oil-water-soap mixture _____

_____

_____

Hard water reactions:

(5) Results of test with $CaCl_2$ solution _____

_____

(6) $MgCl_2$ solution _____

_____

(7) $FeCl_3$ solution _____

_____

(8) Behavior of soap-alcohol solution with phenolphthalein

_____

_____

POST-LAB QUESTIONS

1. Why is it necessary to remove unreacted sodium hydroxide from the soap before using it on the skin?

    _____

2. What is "hard" water? What effect does it have on the cleansing action of soap?

    _____

3. Suppose in this experiment that some students used cotton seed oil, some olive oil and some corn oil. Would their results be similar? Explain.

    _____

4. Think of and state one difference between the soap you made and a bar of commercially available soap.

    _____

32 Electrochemical Cells

OBJECTIVE

To learn how oxidation-reduction reactions can be used to convert chemical energy into electrical energy and to construct working electrochemical cells.

APPARATUS AND CHEMICALS

one 24-well culture plate
seven Beral thin stem pipettes
filter papers and scissors
pair of electrodes: Cu/Zn, Cu/Pb, Cu/Mg,
  Cu/Ag, Cu/Fe
one lemon

0.1 solutions of $Cu(NO_3)_2$, $Zn(NO_3)_2$, $Mg(NO_3)_2$,
  $AgNO_3$, $FeSO_4$, KCL
steel wool
millivolt meter
two wires with alligator clips

SAFETY CONSIDERATIONS

Safety goggles should always be worn in a chemistry laboratory. No other special safety precautions are needed for this experiment since none of the materials used in this experiment are hazardous.

FACTS TO KNOW

An electrochemical cell (Fig. 32.1) is a device for converting chemical energy into electrical energy. Any oxidation-reduction reaction can serve as the basis for designing an electrochemical cell. Reactions based on the difference in the tendency of metals to be oxidized are often used to design an electrochemical cell. For example, the following spontaneous reaction indicates

$$Zn(s) + Cu(No_3)_2 (aq) \rightarrow Zn(s) + Cu(No_3)_2 (aq) + Cu(s)$$

that zinc is more easily oxidized than copper. The net ionic equation illustrates more clearly the changes in oxidation states

$$Zn(s) + Cu^{2+}(aq) \rightarrow Zn^{2+}(aq) + Cu(s)$$

This reaction can be observed by putting a piece of zinc metal in a solution of copper(II) nitrate and observing the metallic copper depositing on the zinc surface. With time the blue color characteristic of Cu^{2+} ions will fade and the amount of metallic copper will increase. At the same time, the strip of zinc metal is gradually being consumed, indicating zinc atoms are being oxidized to zinc(II) ions.

A working electrochemical cell requires separation of the oxidation part of the reaction from the reduction part of the reaction so that the transfer of electrons occurs through an external wire connecting the two parts. The separate parts are represented by half-reactions:

$$Zn \rightarrow Zn^{2+}(aq) + 2\ e^- \qquad \text{oxidation half-reaction}$$

$$Cu^{2+}(aq) + 2\ e^- \rightarrow Cu(s) \qquad \text{reduction half-reaction}$$

Figure 32.1 shows the design of an electrochemical cell based on the reaction of zinc metal with copper(II) ions. The oxidation half-reaction serves as the anode component and the reduction half-reaction serves as the cathode component of the cell. When the anode and the cathode are connected externally by a wire conductor and the anode compartment and cathode compartment are connected by a conducting electrolyte (called a salt bridge), a circuit is completed for electrons to flow externally and the ions to flow internally. If there is a spontaneous chemical reaction between the electrode components, then there is oxidation at the anode and reduction at the cathode; electrons flow spontaneously from the anode to the cathode.

The tendency of the metal to lose electrons is referred to as its activity. Zinc is more active metal than copper on the basis of the experiment just described. In other words, zinc is a better reducing agent than copper—it loses electrons more easily. The greater the difference in activity of two metals, the larger the voltage measured for the electrochemical cell constructed from their oxidation-reduction reaction. In the present experiment you will measure voltages for a number of electrochemical cells, and on the basis of these voltages, determine the relative order of activity of the metals involved.

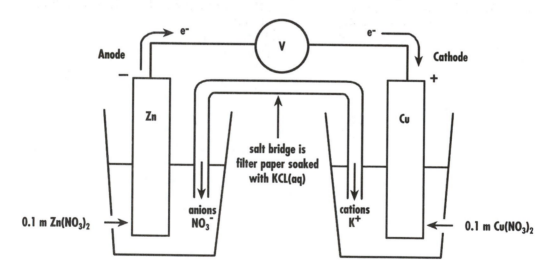

Figure 32-1. Design of an Electrochemical cell.

PROCEDURE

I. Obtain 7 beral pipettes, and label them with a marking pen as follows: KCl, $Cu(NO_3)_2$, $Zn(NO_3)_2$, $Mg(NO_3)_2$, $AgNO_3$, $FeSO_4$. Fill the beral pipettes with the solutions indicated by the label.

II. The salt bridge will be a piece of filter paper about 4.5 cm $\times$ 0.5 cm. Cut ten strips of filter paper this size. Place these on a watch glass and cover with 0.1 M KCl solution.

III. Obtain a 24-well culture plate and measure out 20 drops of 0.1 M $Cu(NO_3)_2$ into each of the wells A-1, A-2, A-3, A-4, A-5 of the 24-well plate. Then take the appropriate Beral pipette to add to wells B-1, B-2, B-3, B-4, B-5: 20 drops of 0.1 M $Zn(NO_3)_2$; 20 drops of 0.1 M $Pb(NO_3)_2$; 20 drops of 0.1 M $Mg(NO_3)_2$; 20 drops of $AgNO_3$; and 20 drops of $AgNO_3$.

IV. Use tweezers to place one of the strips of filter paper between the two wells. Since the filter paper is the salt bridge, make sure the filter paper dips into each well.

V. Polish all the electrode wires with a piece of steel wool.

VI. Take the Cu/Zn electrode pair and place in the adjacent wells (A-1 and B-1) with the Cu electrode in the Cu^{2+} solution and the Zn electrode in the Zn^{2+} solution. Use a set wires with alligator clips to connect the Cu electrode to the positive pole of the voltmeter and the Zn electrode to the negative pole of the voltmeter and record the reading in millivolts (1a). (The Cu is the cathode and the Zn is the anode).

VII. Use the appropriate electrode pairs to measure the voltages for the solution pairs:

Cu/Pb for A-2 and B-2
Cu/Mg for A-3 and B-3
Cu/Ag for A-4 and B-4
Cu/Fe for A-5 and B-5

Record the voltages (2a - 5a). (Note that the Cu is the cathode if you get a positive voltage reading with Cu connected to the positive pole of the voltmeter. If you get a negative voltage, switch the connections to the poles on the voltmeter, and note on the data sheet that the anode is Cu and the cathode is the other electrode material.)

VIII. After measuring the voltages using the 24-well plate, place the electrodes in a lemon and measure and record the voltages (1b - 5b).

PRE-LAB QUESTIONS

1. Given: $2 Al(s) + 3 Cu^{2+}(aq) \rightarrow 2 Al^{3+}(aq) + Cu(s)$

   A. What is oxidized? _____

   B. What is reduced? _____

   C. Which is the more active metal, Al or Cu? _____

2. Refer to Figure 32.1 and design a working electrochemical cell based in the reaction in question (1).

   Why is the strip of filter paper needed in the construction of the electrochemical cells described in the procedure section?

Experiment 32
Data Sheet

Name _____

Date _____

ELECTROCHEMICAL CELLS

	Voltage, millivolts		Anode	Cathode
(1) Cu^{2+}/Cu vs. Zn^{2+}/Zn	(a) _____	(b) _____	_____	_____
(2) Cu^{2+}/Cu vs. Pb^{2+}/Pb	(a) _____	(b) _____	_____	_____
(3) Cu^{2+}/Cu vs. Mg^{2+}/Mg	(a) _____	(b) _____	_____	_____
(4) Cu^{2+}/Cu vs. Ag^{2+}/Ag	(a) _____	(b) _____	_____	_____
(5) Cu^{2+}/Cu vs. Fe^{2+}/Fe	(a) _____	(b) _____	_____	_____

QUESTIONS

I. On the basis of the data above write balanced net ionic oxidation-reduction reactions for:

(1)

(2)

(3)

(4)

(5)

II. Give the order of activity of the metals for the metals used in this experiment based on your voltage measurements for the electrochemical cells. List the most active metal first

III. How do your values measured in the lemon compare to those measured in the 24-well plate? Which are generally larger?

IV. Why does a lemon or other fruit work as part of an electrochemical cell?

POST-LAB QUESTIONS

1. A student forgot to polish the electrode wires with steel wool before using them. the zinc electrode appeared dull because of an oxide coating. Would the measured voltage be lower or higher than expected? Explain.

2. What is the oxidizing agent and the reducing agent in each of the net ionic equations you have listed in Question I of the experiment?

   (1) Oxidizing agent _____   reducing agent _____

   (2) Oxidizing agent _____   reducing agent _____

   (3) Oxidizing agent _____   reducing agent _____

   (4) Oxidizing agent _____   reducing agent _____

   (5) Oxidizing agent _____   reducing agent _____

3. A. From the order of activity of metals you determined in the experiment (Question II), write the net ionic equation for the combination that would give the largest voltage.

   B. Use the net ionic equation in (A) to design a working electrochemical cell.

33 Ion Exchange

OBJECTIVE

To learn the principles by which ion exchange is done
To measure the exchange capacity of a cation-exchange resin

APPARATUS AND CHEMICALS

50 mL buret	cation exchange resin
200 mL volumetric flask	1.5 M NaCl solution
10 mL pipette	3.0 M H_2SO_4 solution
250 mL Erlenmeyer flask	standard NaOH solution (0.1 M)
50 mL beaker	phenolphthalein indicator solution
analytical balance	glass wool

SAFETY CONSIDERATIONS

A burette is easy to break because of its shape, and is consequently, a potential hazard. Think about careful movements as you use this long piece of glass.

The acid (H_2SO_4) and base (NaOH) solutions as well as the solution collected from the ion exchange column (an HCl solution) are damaging to soft tissues—particularly the eyes. Wear eye protection at all times.

Do not use your mouth on a pipette. Use a rubber suction bulb. It is easy to get corrosive chemicals into your mouth using improper technique with a pipette.

Be careful working with the glass wool. Small slivers can get into your fingers. If gloves are available, wear them when using glass wool.

FACTS TO KNOW

Many of the substances that are soluble in water are ionic in nature. The solubility of an ionic substance is largely a consequence of the *charges* the ions themselves carry, in combination with the *highly polar character* of the water molecule. When these ionic substances dissolve, the ions dissociate from one another and are largely independent of one another in the solution—that is, they have their own characteristics that generally do not change, regardless of what the balancing ions are (Na_{aq}^+ is the same, whether it comes from dissolving NaCl, Na_2SO_4, or NaOH).

The independent nature of the ions in an aqueous solution allows the manipulation of the ionic character of the solution. Precipitation reactions are one good example of this type of manipulation, where one ion can be replaced by another ion by causing the first to form a solid which separates from the solution due to solubility considerations. Consider a solution of NaCl in water—it contains Na^+ and Cl^- ions. Adding a stoichiometric quantity of (dissolved) $AgNO_3$ (which is highly soluble in water) will cause the precipitation of AgCl, which is NOT soluble in water, leaving the Na^+ and NO_3^- ions (effectively a solution of the salt $NaNO_3$ (Equation 33.1). The net effect has been to exchange the NO_3^- ions for the Cl^- ions.

$$Na_{(aq)}^+ + Cl_{(aq)}^- + Ag_{(aq)}^+ + NO^3{}_{(aq)}^- \rightarrow Na_{(aq)}^+ + NO_{3(aq)}^- + AgCl_{(s)} \qquad 33.1$$

Ion exchange such as takes place in a precipitation reaction can be useful for many purposes, including *synthesis reactions* (the above example could be a synthesis for either $NaNO_3$ or AgCl, since the two products are easily separated—the AgCl by simple filtration and the $NaNO_3$ by careful evaporation of the filtrate to leave the solid salt), *qualitative analysis* (adding $AgNO_3$ to a solution will NOT result in a precipitate if Cl^- is not present, so the appearance of a precipitate is an indication of the presence of chloride in the solution), and *quantitative analysis* (adding an excess of $AgNO_3$ will allow precipitation of ALL of the chloride as AgCl, so determining the mass of precipitate formed will determine the amount of chloride initially present).

Ion exchange processes are used in many homes on an everyday, ongoing basis. Dissolved minerals in hard water are ionic substances. These would include the cations Ca^{2+}, Fe^{3+} and Mg^{2+}, and anions HCO_3^-, Cl^-, NO_3^-, and SO_4^{2-}. Many people choose to have water softeners installed in their homes. These are simply ion exchange units (like the units used in this experiment, only much larger) that exchange sodium ions (Na^+) for any other cations present. The anions are not affected. This "softens" the water because sodium compounds are the most soluble of all ionic salts and thus insoluble salts and residues are not formed as easily. Calcium, iron, and magnesium form insoluble precipitates with soap molecules, for example, producing a grey-white "scum" that reduces the cleaning power of soap, and calcium can react with bicarbonate in hot water to form insoluble $CaCO_3$ (boiler scale that accumulates in cooking dishes and boilers) (equation 33.2).

$$Ca_{(aq)}^{2+} + 2\,HCO_{3(aq)}^- \rightarrow CaCO_{3(s)} + H_2O + CO_{2(g)} \qquad 33.2$$

Sodium does not form a precipitate with soap, as the soap "molecule" is actually a sodium salt itself! Neither does sodium form a precipitate with HCO_3^- in hot water.

A very convenient method of ion exchange involves the use of materials that are ionic, but which consist of one ion that has no tendency whatsoever to dissolve (generally due to its very large size) and a second ion which easily dissociates into solution. Imagine a very large molecule (one with a molecular weight > 30,000 or as high as 500,000). This would not be very soluble in water. However, if one part of that molecule were an ionic portion containing, for example, a sodium ion, the sodium ion could easily be dispersed into the water if another ion replaced it in the larger molecule. A portion of a molecule such as this, called a polymer, is shown in Figure 33-1.

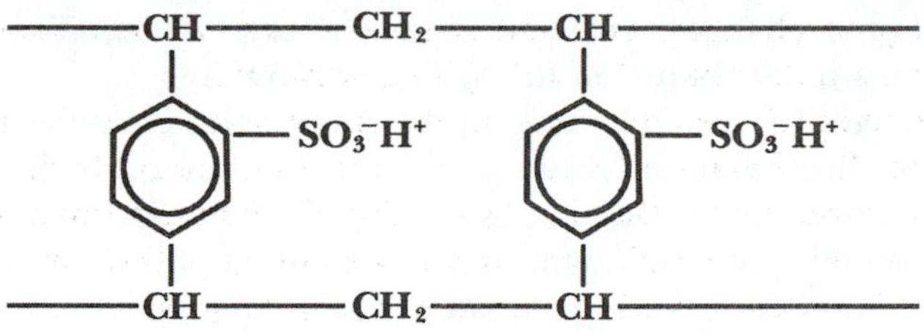

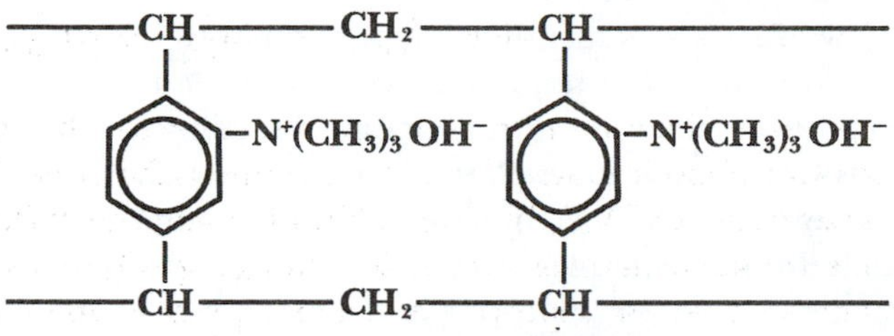

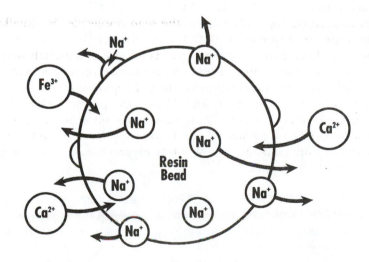

Anion exchange resin structure

Figure 33-1. Portion of a ion-exchange polymer.

    The entire polymeric molecule is simply this structure repeated thousands of times. The physical nature of the polymer is that it is a rubbery plastic type of material that is manufactured as small beads. This would be like having a small rubber ball with sodium ions on the surface. The sodium ions could not just leave the ball and go into the water (this would give an excess of dissolved cations in solution, which cannot happen because the solution must remain electrically neutral). If other cations already present in the water (from some other dissolved ionic materials) had a stronger attraction for the "rubber-ball anion," they would *replace* the sodium ions on the surface of the ball and the sodium ions would replace the other cations in solution (Fig. 33-2). By removing the rubber ball (simple filtration) one would remove the original cations in the solution and replace them with the sodium ions that had originally been a part of the rubber ball.

Figure 33-2. Cation exchange resin.

There are both cation-exchange resins and anion-exchange resins. The *cation-exchange resins* generally have either H^+ ions or Na^+ ions on the surface to exchange with ions in the solution. *Anion-exchange resins* generally have Cl^- or OH^- ions on the surface to exchange with anions in the solution (see Fig.33-1). When you use deionized water, you are using water that has been passed first over a cation-exchange resin that exchanges all cations for H^+ ions, then over an anion-exchange resin that exchanges all anions for OH^- ions. Since the water is electrically neutral (total cation charge equals total anion charge) the number of H^+ ions produced exactly equals the number of OH^- ions produced. These two ions neutralize each other to produce water, and there are no more ions left in the water. Dissolved ionic substances in natural water supplies are called minerals, so deionized water is sometimes referred to as demineralized water. It is simply water where all cations have "become" H^+ and all anions have "become" OH^-, and they have reacted together (no more ions!).

It is important to know exactly how many ions a given quantity of resin can exchange. This quantity is known as the *exchange capacity* of the resin. In a home for example, this information is necessary to allow a determination of how frequently to regenerate the resin in a water softener, based on the hardness of the water. Today in lab you will use a cation-exchange resin. The resin you will use exchanges any metal cations (Na^+, Ca^{2+}, etc.) with H^+ ions from the resin. This is an equilibrium that can be pictured with the following examples:

$$RESIN^-H^+_{(s)} + Na^+_{(aq)} \Leftrightarrow RESIN^-Na^+_{(s)} + H^+_{(aq)} \qquad 33.3$$

$$2\ RESIN^-H^+_{(s)} + Ca^{2+}_{(aq)} \Leftrightarrow (RESIN^-)_2Ca^{2+}_{(s)} + 2\ H^+_{(aq)} \qquad 33.4$$

The resin can be regenerated (put back in the "acid" form) by forcing the equilibria above to go back to the left. This is accomplished by adding a very large excess of acid so that the resin is "swamped" with H^+ ions. We will use the resin in this experiment once, and then regenerate it for the next class.

PROCEDURE

The first part of the experiment is to determine the exchange capacity of the cation-exchange resin. This is a measure of how many ions one gram of the resin can exchange, and has the units of meq/gram were meq stands for milliequivalents of H^+.

> **PREPARATION OF CATION-EXCHANGE COLUMN:** Into a clean and dry 50 mL beaker weigh out approximately 8 grams of cation-exchange resin on the analytical balance, recording the exact weight (1 & 2). Using your squeeze bottle, fill the beaker about half full with deionized water. Assemble a clean burette and put a very small amount of glass wool in the bottom of the burette (your instructor will demonstrate this). Carefully transfer the weighed resin into the burette and wash the sides of the burette with deionized water until all of the beads have collected in a column in the bottom of the burette. Draw the water level down to about 1/8 inch above the top of the beads. *From this point on in the experiment, never let the liquid level fall below the top of the beads!* Doing so would allow air bubbles to be trapped in the column of beads and would make the ion exchange process with the liquid less efficient.

> **ION EXCHANGE BETWEEN SOLUTION AND RESIN:** Using a 10 mL pipette, allow 10 mL of 1.5 M NaCl solution to run down the inside of the burette and collect on top of the resin column (this reagent is being added in excess). Place a 200 mL volumetric flask under the burette, open the stopcock, and let the solution drip slowly into the flask. *When the liquid level has dropped to 1/8 inch above the top of the resin column, close the stopcock*—then add 10 ml of deionized water (measured with a graduated cylinder) carefully to the burette. Let this solution drip into the flask, *again not allowing the liquid to drop below the top of the column.* Repeat this process 3 more times (until you have washed a total of 40 mL of deionized water over the column). All solution coming through the column should be collected in the volumetric flask. Dilute the solution in the flask to the mark etched on the neck and mix well. This solution contains all of the exchangeable H^+ ions from the resin you weighed out, in an accurately known volume (recall that the volume of the volumetric flask is accurate to 200.0 mL). The solution also contains the Cl^- ions from the original NaCl

solution, as well as the excess Na^+ ions that the resin could not exchange. It is now a solution of HCl and NaCl, with the HCl being present as a result of ion exchange. By determining the concentration of HCl in this solution, one can determine the exchange capacity of the resin. Set this solution aside for analysis after the next portion of the experiment.

REGENERATION OF THE RESIN: The resin must now be "regenerated" (put back into the "acid" form which has exchangeable H^+ ions). Add 25 mL of 3 M H_2SO_4 solution to the burette and let the solution drip into a waste beaker (this solution will be discarded). *When the liquid level has fallen to 1/8 inch above the resin, close the stopcock* and add another 25 mL of 3 M H_2SO_4. Repeat this process with one additional portion of 3 M H_2SO_4 (75 mL total used). Finally, wash 50 mL of deionized water over the resin beads. The washings may be discarded in the sink and flushed with tapwater. The resin can then be washed out of your burette into a collection beaker. Your instructor will help you with this. The burette should be thoroughly rinsed with distilled water, as it is to be used in the next part of the experiment.

DETERMINATION OF THE EXCHANGE CAPACITY OF THE RESIN: The concentration of the HCl solution can be determined by titrating the acid against a solution of standard base. Obtain 100 mL of the standard NaOH solution (your instructor will supply the exact normality) in a clean, dry beaker. Rinse the burette with several small portions of this base solution (discarding the washings) and fill the burette with the NaOH solution. Eliminate air bubbles from the tip and take an initial reading (8A).

Using a 20 mL pipette (20.00 mL), transfer a 20 mL aliquot of the HCl/NaCl solution to a 250 or 300 mL Erlenmeyer flask. Your instructor will demonstrate the use of a pipette. Add about 25–50 mL of deionized water, 2 drops of phenolphthalein, and titrate to the first permanent pink endpoint. Take a final burette reading (8B). The volume of base used is the difference between the final and initial readings (8C). Repeat this process with at least 2 more 20 mL aliquots of the unknown solution (minimum of 3 titrations—you may do more if time permits) (9A, 9B, 9C → 12A, 12B, 12C). Refill the burette if it appears that you will have insufficient base for any titration. You probably can get 2 titrations without refilling the burette. Calculate the normality of the HCl solution (8D → 12D) *for each titration* as shown below, then calculate the average normality for all titrations (13). Using this value, the total milliequivalents of acid exchanged by the resin can be determined (14). The exchange capacity of the resin is the total milliequivalents of acid exchanged divided by the total mass of resin used in the column (15).

CALCULATIONS

1. Determine the normality of the HCl solution for each titration (8D → 12D).

$$mL_{(base)} \times N_{(base)} = mL_{(acid)} \times N_{(acid)}$$

2. Determine the average normality for all titrations (13).

3. Determine the total meq of acid exchanged (14). (N is meq/mL)

$$N_{(acid)} \times 200 \text{ mL} = \text{total meq exchanged*}$$

*Remember that all of the solution collected was diluted to 200.0 mL.

Determine exchange capacity of resin (15).

$$\text{Exchange capacity} = (\text{total meq exchanged})/(\text{mass of resin})$$

PRE-LAB QUESTIONS

1. In this experiment, an ion exchange resin is used. What are the two types of cations that cation-exchange resins generally exchange for cations in solution?

    _____

    _____

2. What are the cations in hard water that are responsible for the hardness of the water?

    _____

    _____

3. What solution is initially added to the ion exchange resin in this experiment, what volume of the solution is added, and how is it added?

    _____

    _____

4. How is the quantity of exchanged ions determined?

    _____

    _____

    _____

    _____

Experiment 33
Data Sheet

Name _____

Date _____

ION EXCHANGE

Mass of 50 mL beaker _____ (1) (If tared, write "tared.")

Mass of beaker + cation-exchange resin _____ (2)
    (or mass of tared resin)

Mass of resin _____ (3)

Volume of volumetric flask used in the experiment _____ (4)

Volume of NaCl solution pipetted onto resin _____ (5)

Concentration of NaCl solution used in experiment _____ (6)

Volume of pipette used for the diluted solution _____ (7)

Burette readings for titrations:

Trial number	Initial reading	Final reading	Volume used	Normality
1	_____ (8A)	_____ (8B)	_____ (8C)	_____ (8D)
2	_____ (9A)	_____ (9B)	_____ (9C)	_____ (9D)
3	_____ (10A)	_____ (10B)	_____ (10C)	_____ (10D)
4	_____ (11A)	_____ (11B)	_____ (11C)	_____ (11D)
5	_____ (12A)	_____ (12B)	_____ (12C)	v12D)

Average normality of HCl _____ (13)

Total Milliequivalents of acid exchanged _____ (14)

Exchange capacity of ion exchange resin _____ (15)

POST-LAB QUESTIONS

1. What is the reason for using the volumetric flask to collect the solution eluting from the ion exchange column?

   _____

   _____

   _____

2. What procedure insures that all of the exchanged ions have been washed off the ion exchange column for analysis.

   _____

   _____

   _____

3. If the ion exchange column had gotten air bubbles entrapped in the beads, what effect would this have had on the analysis, and WHY?

   _____

   _____

   _____

4. Describe how one makes deionized or demineralized water, beginning with tapwater which contains dissolved minerals. (How does one make all of the ions "disappear.")

   _____

   _____

   _____

   _____

34 Carbohydrates: The Identification of an Unknown

OBJECTIVE

To become familiar with some of the characteristic reactions of carbohydrates.
To identify an unknown carbohydrate.

APPARATUS AND CHEMICALS

small test tubes (4)
8″ test tube
400-mL beaker
glass plate
ring stand
iron ring
Bunsen burner
iodine solution
water bath at 37°C

Seliwanoff reagent
Benedict's solution
fructose
sucrose
cellulose
starch
lactose
yeast suspension

SAFETY CONSIDERATIONS

    Wear eye protection because of possible shattering of dropped glassware and because acids are used in this experiment.

    Seliwanoff reagent is resorcinol dissolved in 4 M (molar) hydrochloric acid. The acid is corrosive, that is, it burns human tissue. If any of this solution gets on your skin, wash the solution off immediately with water. Avoid breathing the vapors of the solution.

    Protect your hair, clothing, and flammable materials from the open flame of the burner. Make sure the beaker is well supported when boiling water so the hot water will not spill on you.

FACTS TO KNOW

The carbohydrates ("hydrates of carbon") are one of the important classes of organic compounds that are composed of the three elements of carbon, hydrogen, and oxygen. These three elements are bonded in such a way that a number of alcohol groups (−OH) occur along with an aldehyde (−CHO) or ketone (−CO−) group. A simple sugar such as glucose is an example of a carbohydrate with some of these structures. Glucose exists almost entirely in the cyclic form.

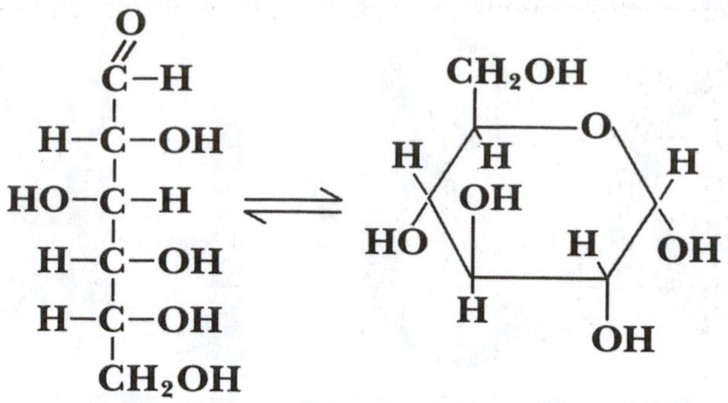

Two forms of glucose

The most important simple carbohydrate is glucose. It is found as a major component of other carbohydrates. For example, maltose is a carbohydrate composed of two glucose units. Lactose (milk sugar) contains one glucose unit and one galactose unit:

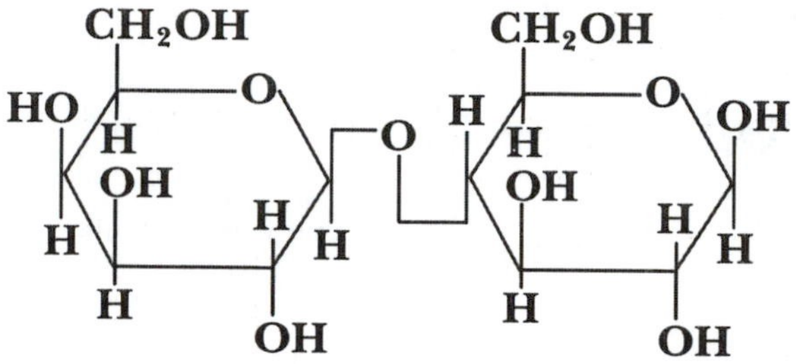

Galactose unit   Glucose unit

Lactose

Sucrose contains one glucose unit and one fructose unit:

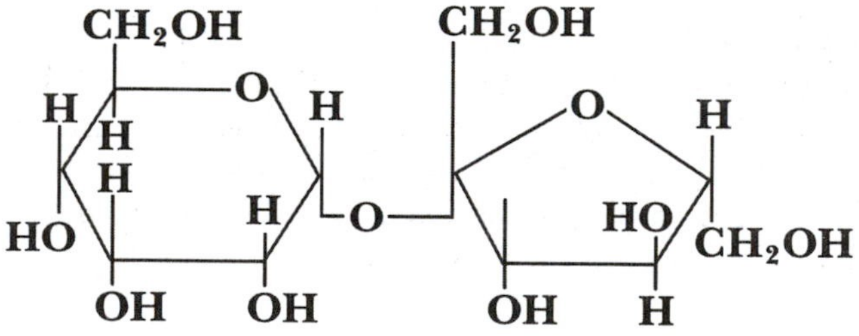

Glucose unit   Fructose unit

Sucrose

Starch and cellulose are carbohydrates that contain a large number of glucose units:

Structures of starch (amylose)

Cellulose has basically the same structure as starch except the orientation of the bond that connects the glucose units is different.

All the carbohydrates—fructose, sucrose, cellulose, starch, and lactose—have unique chemical properties that allow them to be identified. Some react similarly to the same test; however, by using a number of tests each of the sugars may be identified by its unique set of reactions. Ketoses, sugars which contain a ketone group,

$$R - \overset{\overset{\displaystyle O}{\|}}{C} - R$$

$(R - C - R)$, often differ in their chemical properties from the aldoses, sugars which contain an

aldehyde group $(R - \overset{\overset{\displaystyle O}{\|}}{C} - H)$. The following test will be used:

I. FERMENTATION   The common hexoses (six-carbon sugars), with the exception of galactose, are fermented to carbon dioxide and ethanol by the action of yeast. This is accomplished with the aid of the enzyme zymase.

$$C_6H_{12}O_6 \xrightarrow{\text{Zymase}} 2\ CH_3CH_2OH + 2\ CO_2$$

A common hexose   ethanol

Yeast contains other enzymes, such as maltase and sucrase, which split maltose and sucrose, respectively, into the simpler glucose (and fructose) units. The glucose then may break down to ethanol and carbon dioxide. Of the carbohydrates listed previously, cellulose and lactose are the only common ones which resist fermentation by ordinary yeast.

II. BENEDICT'S TEST   Any carbohydrate with a potentially free carbonyl group

$$-\overset{\overset{\displaystyle O}{\|}}{C}-$$

$(- C -)$ will reduce Benedict's solution. The carbonyl groups of carbohydrates such as sucrose, starch, and cellulose are engaged in bonding and will not reduce Benedict's solution.

III. SELIWANOFF TEST   When a ketose is heated with a strong mineral acid, hydroxymethyl furfural is formed. This compound forms a red complex with the organic compound, resorcinol. The aldoses will give the red color more slowly.

Fructose (a ketose)   **Hydroxymethyl furfural**

IV. IODINE TEST Starch forms a dark blue complex when treated with iodine. Cellulose and the simple sugars produce no color change at all. A single drop of the iodine solution gives an unmistakable test for starch.

PROCEDURE

The table on the data sheet indicates + or − for positive or negative tests for fructose, sucrose, cellulose, starch, and lactose. Run the following tests on your unknown and record a + or − in the table.

I. FERMENTATION In a large test tube place 4 mL of yeast suspension and about 30 mL to 40 mL of the unknown carbohydrate solution. Mix well. Place a small test tube upside down in the larger test tube (Fig. 34-1a). Place the palm of your hand over the mouth of the large test tube and invert so that the smaller test tube is filled (Fig. 34-1b). Return the large test tube to the upright position. The smaller tube should be filled completely and resting inverted within the solution (Fig. 34-1c). Place the apparatus in a water bath maintained at 37°C and see if bubbles of CO_2 collect within the smaller tube. If none appear within two hours, it can be assumed that the solution is not acted upon by yeast.

II. BENEDICT'S TEST Place 2 mL of Benedict's solution in a small test tube and heat in a gently boiling water bath for 2 or 3 minutes. Add 10 drops of the solution to be tested, mix thoroughly, and allow it to heat an additional 5 minutes. A color change from blue to green to yellow and finally to brick red (Cu_2O) indicates that the carbohydrate is a reducing sugar. The rate of reaction is very important. Fructose reacts very rapidly, glucose and galactose more slowly. Even starch and cellulose will give weak tests if the solution is heated extensively.

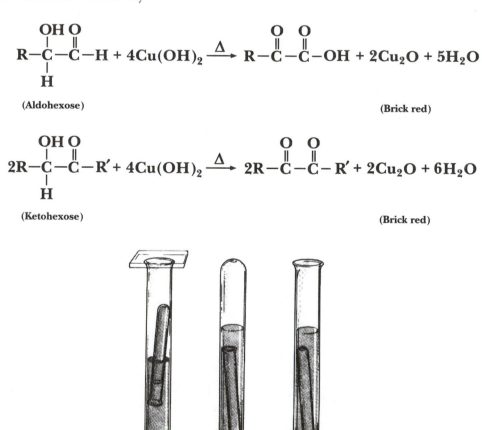

Figure 34-1. Use of test tubes for fermentation reaction.

III. SELIWANOFF TEST   To 5 mL of freshly prepared Seliwanoff reagent, add 10 drops of the unknown solution and boil for *less than* 20 seconds in a water bath. (*BE CAREFUL!!* This is a strongly acidic solution.) A red color at this time is indicative of a ketohexose. If the red color develops after 20 seconds, it is likely that the unknown is an aldohexose (glucose).

IV. IODINE TEST   Take 2 mL of the unknown solution and add 2 drops of iodine solution. A deep blue indicates the presence of starch.

PRE-LAB QUESTIONS

1. If your skin or eyes contact the resorcinol solution or any of the other solutions, what should be your immediate action?

    _____

    _____

    _____

2. Of the naturally occurring carbohydrates studied in this experiment, lactose, sucrose, starch, and cellulose, what is the hexose common to all?

    _____

    _____

    _____

3. Give examples of two simple sugars found in this experiment.

    _____

    _____

    _____

Experiment 34
Data Sheet

Name _____

Date _____

CARBOHYDRATES: THE IDENTIFICATION OF AN UNKNOWN

	Fermentation		Benedict's Solution		Seliwanoff Reagent		Iodine	
	Known	Unknown	Known	Unknown	Known	Unknown	Known	Unknown
Fructose	+		+		+		−	
Sucrose	+		−		+		−	
Cellulose	−		−		−		−	
Starch	+		−		−		+	
Lactose	−		+		−		−	

My unknown carbohydrate is _____

POST-LAB QUESTIONS

1. In the fermentation of a simple glucose, what gas is given off?

    _____

2. Many carbohydrates are naturally occurring. Most have nutritional value which we consume. Starch and cellulose are two similar, but not identical, carbohydrates which are found in large abundance in our environment. Starch is readily broken down by the body to produce energy while cellulose (found in the wood material) is not. Give a possible reason for this.

    _____

    _____

    _____

3. What determines whether or not a carbohydrate will give a positive Benedict's Test?

    _____

    _____

    _____

4. Which of the carbohydrates studied in this experiment would be consistent with the following data?
    Benedict's Test-Positive   Iodine Test-Negative
    Seliwanoff Test-Negative   Fermentation-Negative

    _____

    _____

    _____

35 The Enzymatic Breakdown of Starch

OBJECTIVE

To examine the enzymatic breakdown of starch.
To determine the "ptyalin" number of a sample of saliva.

APPARATUS AND CHEMICALS

4-inch test tubes (11)
ring stand
iron ring
Bunsen burner
10-mL graduated cylinder
25-mL graduated cylinder

100-mL graduated cylinder
400-mL beaker
wax pencil
110°C thermometer
2% starch solution
0.01 M iodine solution

SAFETY CONSIDERATIONS

Wear safety glasses to protect against shattered glass and spilled chemicals.

The Bunsen burner is used in this experiment. Care should be exercised in igniting and using the burner. Do not forget to turn the gas off after use.

Do not touch hot glassware or hot metal ring stands.

FACTS TO KNOW

In order to obtain the desired products from a chemical reaction, the chemist sometimes has to employ very drastic conditions. This may include making the reacting solution very acidic or basic, or possibly using very high temperatures. An alternative to using drastic conditions is to select a substance to act as a catalyst, i.e., a compound that will speed up the rate of the reaction without being consumed in the overall process. A typical example would be the hydrolysis of sucrose to form glucose and fructose.

$$C_{12}H_{22}O_{11} + H_2O = C_6H_{12}O_6 + C_6H_{12}O_6$$

Sucrose   Glucose   Fructose

If sucrose were placed in water by itself, the sugar would take a very long time to break up. However, if about 20% acid is added, the acid catalyst will speed up the reaction and allow the sucrose to be hydrolyzed in about 30 minutes at 50°C. Although this is a great improvement over no catalyst, the severe conditions do not cause immediate hydrolysis of sucrose.

Another possibility is the selection of an enzyme as catalyst. **Enzymes** are the biochemicals that catalyze the synthesis and breakdown of biochemical molecules. Under physiological temperatures (37°C) the enzyme, sucrase, will catalyze the breakdown of sucrose in a matter of seconds.

This experiment involves a similar process, the breakdown of *starch* units into simpler carbohydrates. The effective enzyme is called ptyalin and is found in human saliva. The activity, or rate of action, of the ptyalin varies widely from person to person. In this experiment the individual activity will be determined by calculation of the **"ptyalin number,"** defined as **the number of milliliters of 1% starch solution that can be digested in 30 minutes by 1 mL of saliva.** The experimental procedure is based on the fact that a solution of starch will turn dark blue when mixed with a few drops of iodine solution. However, when the starch becomes hydrolyzed to the smaller carbohydrate fragments, the blue color disappears. Therefore, when the starch-enzyme solution goes to the achromic (colorless) point, the starch will be considered hydrolyzed.

Amylose structure of starch   **Maltose**

$$(C_6H_{10}O_5)_n + xHOH \rightarrow C_6H_{12}O_6 + \text{Other fragments}$$

Sucrose   Smaller molecular weight carbohydrates
(blue in iodine solution)   (colorless in iodine solution)

PROCEDURE

Collect about 10 mL of saliva in a clean test tube. (You can easily spare this much since you normally secrete approximately 1.5 quarts daily.) If you have eaten recently, it may be best to filter the saliva through a gauze. (**Note:** *While you are doing this you should begin heating a water bath. A 400-mL beaker would be best for the experiment. The bath should be maintained at 37°C ± 2°C.*) For your standard solutions, thoroughly mix 2 mL of the collected saliva with 18 mL of distilled water (stock solution #1; a 10% solution.) For stock solution #2, thoroughly mix 1 mL of saliva with 99 mL of distilled water (a 1% solution).

Number ten clean test tubes, 1 through 10, with a wax pencil and prepare the ten solutions as indicated in the table on the data sheet. Now add 3 mL of 2% starch solution to each of the ten test tubes, mix thoroughly, and when *all* have been prepared, place them into the bath simultaneously.

After exactly 30 minutes, add 2 drops of I_2/KI reagent to each test tube. Record the resultant color in the table. ("Clear" is not a color; use "colorless" or "achromic.") Determine the minimum saliva concentration needed to reach the achromic point and calculate the ptyalin number.

Ptyalin number = _____ maximum number of mL of 1% starch digested (hydrolyzed) per mL of saliva in 30 minutes.

EXAMPLE CALCULATION

Assume that the most dilute concentration of saliva that was achromic was 1%, test tube #7.

$$.01 \times 3 \text{ mL} = .03 \text{ mL of saliva in 3 mL of 1\% solution.}$$

The 3 mL of 2% starch is the same as 6 mL of 1% starch solution. So, 6 mL of 1% starch solution required .03 mL of saliva to reach the achromic point. The ptyalin number is then given by

$$\text{Ptyalin number} = \frac{6 \text{ mL starch}}{.03 \text{ mL saliva}} = 200$$

Answer questions 1 through 3.

PRE-LAB QUESTIONS

1. Define "catalyst" and give one example of a general class of compounds frequently used as catalysts.

   _____

2. What is a major difference between an enzyme and an ordinary catalyst?

   _____

3. If the ptyalin number is to be measured with a 1% starch solution, why is a 2% starch solution employed in this experiment?

   _____

   _____

   _____

Experiment 35
Data Sheet

Name _____

Date _____

THE ENZYMATIC BREAKDOWN OF STARCH

Tube No.		Color with I_2/KI
1	3 mL pure saliva :100.0%	
2	2 mL of pure saliva + 1 mL of water :66.7%	
3	1 mL of pure saliva + 2 mL of water :33.3%	
4	3 mL of 10% saliva :10.0%	
5	2 mL of 10% saliva + 1 mL of water :6.7%	
6	1 mL of 10% saliva + 2 mL of water :3.3%	
7	3 mL of 1% saliva :1.0%	
8	2 mL of 1% saliva + 1 mL of water :0.7%	
9	1 mL of 1% saliva + 2 mL of water :0.3%	
10	3 mL of distilled water :0.0%	

Ptyalin number = _____

POST-LAB QUESTIONS

1. Why does a cracker taste sweet after it has been in your mouth for a while?

    _____

2. The complete hydrolysis of starch yields what simple sugar?

    _____

3. If a person's ptyalin number is 200 and another person's ptyalin number is 150, which person can digest larger amounts of starch more effectively?

36 Red Cabbage as a pH Indicator

OBJECTIVE

To measure the pH of household items with a natural indicator and to test the effect of dilution upon pH.

APPARATUS AND CHEMICALS

400-mL beakers (2)
test tube rack
about 20 test tubes
eyedroppers
labels
conical funnel
filter paper
5-mL graduated cylinder
250-mL beakers (2)
stirring rod
100-mL graduated cylinder
Bunsen burner
ring
ring stand

wire gauze
red cabbage
about 7 solutions spaced between
  pH = 2 and pH = 10
indicator strips or universal
  indicator paper
0.20 M acetic acid
0.20 M sodium acetate
0.10 M hydrochloric acid
consumer products such as shampoos,
  antacids, cleansers, colorless
  soft drinks, vinegar, soaps
0.10 M sodium hydroxide

SAFETY CONSIDERATIONS

 Wear eye protection to prevent eye injury from shattered glass and spilled acids and bases.

 Keep clothing and hair away from open flame of the burner. The beaker used to boil the cabbage should be well supported. Make sure the stand and beaker do not tip over.

FACTS TO KNOW

The concept of pH is encountered in advertising material such as in the advertising of shampoos, antacids, dishwashing liquids, and other consumer products. The control of pH is important in biochemical processes, analytical procedures, and industrial processes.

Concentrations of solutions, including those of acids and bases, are often expressed in **molarity,** which is the number of moles of solute per liter of solution. For example, a 0.1 M solution of hydrochloric acid contains 0.1 moles of HCl(aq) per liter. Since HCl(aq) is 100% ionized, the concentration of H_3O^+ is also 0.1 M.

$$HCl(aq) + H_2O(l) \rightarrow H_3O^+(aq) + Cl^-(aq)$$

Pure water is neutral because it contains equal numbers of hydronium, H_3O^+, and hydroxide ions, OH^-. Although pure water ionizes only slightly, the actual concentration of hydronium ions and hydroxide ions at 25 °C is 1.0×10^{-7} M. The product of the molarity of the hydronium ions and hydroxide ions in pure water is $(1.0 \times 10^{-7})(1.0 \times 10^{-7}) = 1.0 \times 10^{-14}$. In aqueous solutions the product of the molar concentrations of H_3O^+ and OH^- is always 1.0×10^{-14}. Thus, if one is known, the other can be calculated from this expression. For example, a solution with a H_3O^+ concentration of 1.0×10^{-2} molar has an OH^- concentration of 1.0×10^{-12} molar. Danish chemist S.P.L. Sørensen proposed in 1909 that these exponents be used as a measure of acidity.

Sørensen's scale came to be known as the pH scale, from the French *pouvoir hydrogene*, which means hydrogen power. pH is defined as the negative logarithm of the hydronium ion concentration.

$$pH = -\log [H_3O^-]$$

where $[H_3O^+]$ is the molarity (M) of hydronium ions, which is the number of moles of hydronium ion per liter of solution. Some comparable values of hydronium ion concentrations and pH are shown in Table 36-1.

TABLE 36-1   RELATIONSHIPS AMONG [H_3O^+], PH, ACIDITY AND BASICITY OF AQUEOUS SOLUTIONS AT 25°

[H_3O^+], mole/liter	pH
$10^0 = 1$	0
10^{-1}	1
10^{-2}	2
10^{-3}	3
10^{-4}	4
10^{-5}	5
10^{-6}	6
10^{-7}	7
10^{-8}	8
10^{-9}	9
10^{-10}	10
10^{-11}	11
10^{-12}	12
10^{-13}	13
10^{-14}	14

As can be seen from Table 36-1, the pH concept allows expression of acidity in numbers which are the negative exponents of the actual concentration values. The use of exponents is a common practice when dealing with numerical values which span a vast range. Other examples are the earthquake Richter scale and the magnitude order of stars.

The pH of pure, neutral water (neither acidic nor basic) is 7. Solutions with a pH below 7 are acidic; solutions with a pH above 7 are basic.

Each decrease of one pH unit represents a tenfold increase in acidity, or hydronium ion concentration. For example, the H_3O^+ concentration in a solution with pH 3 is ten times that for a solution with pH 4. The actual concentrations are 1.0×10^{-3} and 1.0×10^{-4}, respectively.

The pH of a solution can be measured by a pH meter and/or by acid-base indicators. The pH meter is an apparatus with electrodes sensitive to hydrogen ions. The instrument measures the small voltage produced by the presence of hydrogen ions and reads out the pH.

Many natural substances are acid-base **indicators.** The most familiar one is litmus, an organic dye extracted from certain lichens, which turns from blue to red in acidic solutions (below pH 7) and from red to blue in basic solutions (about pH 7). Other natural indicators include black tea, beet juice, carrot juice, grape juice, blueberry juice, cherry juice, radishes, red cabbage juice, red onions, rhubarb, tomato leaves and flowers such as the rose, day lily, daisy, blue iris, purple dahlia, purple hollyhock.

Most plant extracts contain a mixture of pigments. Because of this, there is usually no sharp color change with pH, but instead a gradual fading from one color into another over a range of several pH units. However, some extracts, such as red cabbage extract, undergo sharp changes of color at several pH values. The deep purple color of the leaves of red cabbage is caused by water-soluble anthocyanins. Over the pH range of 2 to 11, the different anthocyanins will display several colors and shades of color, making red cabbage extract a "universal indicator."

Red cabbage extract is obtained either by boiling a mixture of red cabbage leaves and water or by using a blender to chop up several shredded cabbage leaves covered with water. The mixture from either method is filtered to obtain the extract.

Acid-base indicators are weak acids and bases. A typical indicator will ionize in solution according to the equation

$$HIn = H^+ + In^-$$

The chemical species, HIn and In⁻, are different colors. When the solution is acidic to the degree the HIn species dominates, the solution will indicate the color of HIn. When the solution is more basic where In⁻ dominates, the solution will indicate the color of In⁻. If the acidity is between these two extremes, the color is mixed because both HIn and In⁻ are present.

An acid-base **buffer** controls the pH of a solution within a narrow range, even when small amounts of acid or base are added. A buffer must contain an acid that react with any added base, and at the same time it must contain a base that can react with any added acid. It is also important that the acid and base components of a buffer solution not react with each other. Buffers usually are mixtures of a weak acid and its conjugate base, or a weak base and its conjugate acid. In this experiment the buffer is acetic acid and sodium acetate. According to Brønsted-Lowry theory,

$$CH_3COOH + H_2O \Leftrightarrow H_3O^+ + CH_3COO^-$$

$$\text{acid} \qquad \text{base} \qquad \begin{array}{c}\text{conjugate}\\\text{acid}\end{array} \qquad \begin{array}{c}\text{conjugate}\\\text{base}\end{array}$$

Acetic acid is a weak acid and the acetate ion is a strong conjugate base. This mixture will work as a buffer because acetic acid will react with added base to produce acetate ion while added acid will react with acetate ion to produce acetic acid. Thus the pH is maintained.

PROCEDURE

Numbers in parentheses refer to entry numbers on the data sheet.

You may use tap water throughout this experiment.

PREPARATION OF RED CABBAGE EXTRACT Boil 10 g of red cabbage (cut into small pieces) in 200 mL of tap water in a 400-mL beaker. Continue to boil the solution for about 10 minutes.

Let the solution cool until you can handle the beaker safely.

Fold a filter paper, set up a funnel and support, and filter the extract. See Figure 39-1.

Save the filtered extract for testing the pH of solutions presented to you.

PREPARATION OF pH STANDARDS Label 7 test tubes with the same pH as the standard solutions prepared for you to use in this experiment.

Add 2 mL of each prepared pH solution to its properly labelled test tube.

Then, add 2 mL of cabbage extract to each test tube. Mix the solutions well.

Describe the color of the cabbage extract in each pH (1).

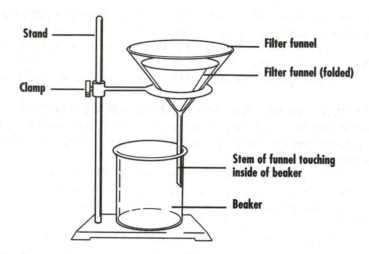

Stand —

Clamp —

Filter funnel

Filter funnel (folded)

Stem of funnel touching inside of beaker

Beaker

Figure 36-1. Set-up for filtering the reaction mixture.

Save these solutions for comparison with cabbage extract in consumer products and in diluted solutions. If the color of a standard changes significantly during the course of the lab, you should remake that standard.

Test each of the standard solutions with a separate strip of universal indicator paper. Record the color produced on the indicator paper (2).

pH OF CONSUMER PRODUCTS Arrange as many test tubes in a test tube rack as you have consumer products to test. Label each test tube with one of the consumer products. Add 2 mL of each consumer product to its properly labelled test tube. Add 2 mL of cabbage extract to each test tube, and mix each solution well. Record the color produced in each solution. Compare with the standard solutions prepared earlier and decide on a pH of each consumer product under the conditions tested (3).

Test each consumer product with a separate strip of universal indicator paper. Consult the standard chart (included with the universal indicator paper) and determine the pH of each consumer product. Record the pH (4).

Arrange the consumer products tested in order from most acidic to most basic (5).

EFFECT OF DILUTION Prepare a buffer solution by mixing 5 mL of the prepared acetic acid solution (0.20 M) with 5 mL of the prepared sodium acetate solution (0.20 M). Use a 5-mL graduated cylinder to measure the volumes. Mix the two solutions thoroughly in a small beaker.

Measure 2 mL of the buffer solution into a test tube. Add 2 mL of the cabbage extract. Mix well. Compare this solution with your cabbage extract standards, and decide on the pH of this solution (6).

Add 2 mL of the buffer solution to a small beaker. Add 100 mL of water and mix the solution well. Measure 2 mL of this solution into a test tube. Add 2 mL of the cabbage extract. Mix well. Compare this solution with your cabbage extract standards and decide on the pH of this solution (7).

Measure 2 mL of the hydrochloric acid solution into a test tube. Add 2 mL of the cabbage extract and mix the solution well. Using the standards, decide on the pH of this solution (8).

Measure 2 mL of the hydrochloric acid solution into a small beaker and add 100 mL of water. Mix the solution well. Measure 2 mL of this solution into a test tube and add 2 mL of the cabbage extract. Mix well and determine the pH of this solution (9).

Measure 2 mL of the sodium hydroxide solution into a test tube. Add 2 mL of the cabbage extract and mix the solution well. Using the standards, decide on the pH of this solution (10).

Measure 2 mL of the sodium hydroxide solution into a small beaker and add 100 mL of water. Mix the solution well. Measure 2 mL of this solution into a test tube and add 2 mL of the cabbage extract. Mix well and determine the pH of this solution by comparison with your standard solutions (11).

Which solution changed its pH less upon dilution, the buffer or the strong acid (and base) solutions (12)?

Of what practical value is dilution on strong acidic and basic solutions (13)?

EFFECT OF A BUFFER UPON ADDITION OF EITHER ACID OR BASE   To 2 mL of the buffer solution add 2 mL of the stock (0.10 M) hydrochloric acid solution and 2 mL of the cabbage extract. Compare with your standard solutions and determine the pH of this solution (14).

To 2 mL of the stock (0.10 M) hydrochloric acid solution add 2 mL of water and 2 mL of the cabbage extract. Determine the pH of this solution (15).

Compare your answers to (14) and (15), and deduce the effect of a buffer upon pH change when acid is added (16).

Obtain answers (17), (18), and (19) by using the sodium hydroxide solution in place of the hydrochloric acid solution (which was used to obtain answers (14), (15), and (16)).

UNKNOWN   Obtain an unknown, which is a solution of unknown pH. Add 2 mL of the unknown to a test tube. Add 2 mL of the cabbage extract and mix the solution well. Use the standards to determine the pH of your unknown. Record the pH (20).

Use the universal indicator paper to measure the pH of your unknown with no cabbage extract present (21).

PRE-LAB QUESTIONS

1. What concentration units are represented by the brackets around $[H_3O^+]$?

   _____

2. What is the pH of a solution that contains 0.001 M hydronium ions?

   _____

3. If the pH of a solution is 6.0, what is the hydronium ion concentration?

   _____

4. Group the following pH values in order of increasing acidity—6, 9, 4, 5, 2.

   _____

5. Which is more basic, a pH of 9 or a pH of 12? What is the magnitude of the difference between these two pH values?

   _____

Experiment 36
Data Sheet

Name _____

Date _____

RED CABBAGE AS A pH INDICATOR

STANDARDS: pH   _____ _____ _____ _____ _____ _____ _____

(1) Cabbage extract
    color   _____ _____ _____ _____ _____ _____ _____

(2) Universal
    indicator color   _____ _____ _____ _____ _____ _____ _____

CONSUMER PRODUCTS: _____ _____ _____ _____

(3) pH with cabbage extract   _____ _____ _____ _____

(4) pH with paper   _____ _____ _____ _____

(5) Order from most acidic to most basic:

_____

DILUTION:

(6) pH of undiluted buffer   _____

(7) pH of diluted buffer   _____

(8) pH of undiluted hydrochloric acid   _____

(9) pH of diluted hydrochloric acid   _____

(10) pH of undiluted sodium hydroxide solution   _____

(11) pH of diluted sodium hydroxide solution   _____

(12) Solution with less change in pH   _____

(13) Practical value of dilution of acids and bases

_____

Red Cabbage as a pH Indicator

BUFFER ACTION ON ADDED ACIDS AND BASES:

(14) pH of buffer and hydrochloric acid   _____

(15) pH of water and hydrochloric acid   _____

(16) How is the pH of a buffer changed upon addition of an acid?

_____

(17) pH of buffer and sodium hydroxide   _____

(18) pH of water and sodium hydroxide   _____

(19) How is the pH of a buffer changed upon the addition of a base?

_____

UNKNOWN

(20) pH with cabbage extract   _____

(21) pH with universal indicator paper   _____

POST-LAB QUESTIONS

1. Write the formula for the components of the buffer system you used and show with chemical equations how the buffer maintained pH when small amounts of HCl(aq) and NaOH(aq) were added.

    _____

    _____

    _____

2. An important buffer system in the blood is one made up of sodium dihydrogen phosphate, NaH_2PO_4, and sodium monohydrogen phosphate, Na_2HPO_4. The buffer reactions are based on the $H_2PO_4^-$ and HPO_4^{2-} system. Identify which species is the acid and which species is the conjugate base.

    _____

3. What is the molarity of H_3O^+ for each of the following solutions?

    A. Solution with a pH of 5.0 _____

    B. Solution with a pH of 1.0 _____

    C. Solution with a pH of 4.0 _____

4. What is the molarity of OH^- in each of the following solutions?

    A. Solution with a pH of 14.0 _____

    B. Solution with a pH of 11.0 _____

    C. Solution with a pH of 13.0 _____

5. Give the pH of the following solutions.

    A. 1.0×10^{-3} M NaOH _____

    B. 1.0×10^{-3} M HCl _____

Chemical Properties of Amino Acids and Proteins

OBJECTIVE

To carry out some reactions on representative amino acids and proteins and to use these reactions to determine which of a limited number of constituents is present in the proteins of egg albumin and gelatin.

APPARATUS AND CHEMICALS

test tubes (4)

test tube rack

400- or 600-mL beaker

Bunsen burner

ring

ring stand

wire gauze

glycine solution

cysteine solution

egg albumin solution

ninhydrin reagent solution (0.1%)

concentrated nitric acid

nitroprusside reagent solution (2%)

ammonium hydroxide

mercuric chloride solution (1%)

copper sulfate solution (0.5%)

sodium hydroxide solution (10%)

concentrated hydrochloric acid solution

mercuric nitrate solution (Millon's reagent)

gelatin solution

SAFETY CONSIDERATIONS

 Wear eye protection to ward against shattered glass and spills of acids and bases.

 Nitric acid (HNO_3), hydrochloric acid (HCl), and sodium hydroxide (NaOH) solutions are corrosive poisons. Mercuric chloride ($HgCl_2$) and mercuric nitrate ($Hg(NO_3)_2$) are metabolic poisons. If any of these solutions get on your skin, wash immediately under running water for several minutes.

 Keep hair and clothing away from the open flame of the burner.

FACTS TO KNOW

Protein molecules build up skin, muscles, hair, the heart, liver, kidneys, and other essential organs. Proteins are composed, for the most part, of long chains of amino acids joined together by peptide linkages. An example of a peptide linkage may be seen in the molecules made up of two amino acid residues:

$$H_2N-CH_2-\overset{O}{\underset{}{C}}-N-CH_2-\overset{O}{\underset{OH}{C}}$$
$$\underset{H}{}$$

Glycylglycine

Peptide linkages

$$H_2N-\overset{H}{\underset{CH_3}{C}}-\overset{O}{\underset{}{C}}-N-CH_2-\overset{O}{\underset{OH}{C}}$$
$$\underset{H}{}$$

Alanylglycine

Because there is still a reactive group at either end of these molecules, they can react with other amino acids to form longer chain molecules:

$$H_2N-\overset{H}{\underset{CH_3}{C}}-\overset{O}{\underset{}{C}}-N-CH_2-\overset{O}{\underset{}{C}}-N-CH_2-\overset{O}{\underset{}{C}}-OH$$

Alanylglycylglycine

Peptide linkages

Phenylalanylalanylglycine

Peptide linkages

The human body builds up the proteins it needs from molecules furnished directly or indirectly by its food. The protein molecules in food that we eat are broken down into their amino acid residues through the process of digestion and are then absorbed. The **essential amino acids** contain molecular structures which the body must obtain from food if it is to synthesize other necessary protein structures.

Amino acids contain both amine groups ($-NH_2$ usually) and carboxylic acid groups $\left(-\overset{O}{\underset{}{C}}-OH\right)$.

Because the amine group is a base and the carboxylic acid group is an acid, they react with each other, and the solid amino acids consist of internal salts, or *dipolar* ions. For example:

$$\underset{^+NH_3}{CH_2}-\overset{O}{\underset{O^-}{C}}$$

Glycine dipolar ions

The reactions of amino acids are characteristic of the particular functional groups present. In the protein itself, both the amine group and the carboxylic acid group are usually tied up in the peptide linkage so they do not exhibit many of their usual reactions. However, some of the reactions of amino acids which may be used to detect amino acids are:

1. THE NINHYDRIN REACTION

Most amino acids are colorless; however, they do react with ninhydrin to give a colored product:

Ninhydrin	Amino acid	Blue complex

This is a characteristic reaction of free amino acids.

2. XANTHOPROTEIC REACTION

Anyone who has carelessly spilled nitric acid on his skin soon observes a pronounced *yellow* color, which is especially noticeable if concentrated nitric acid was used. This is the xanthoproteic reaction, and is due to the nitration of the aromatic rings of *tyrosine, tryptophan,* and *phenylalanine* present in skin proteins.

Tyrosine residue in protein	Nitrated tyrosine residue in protein (yellow)

3. NITROPRUSSIDE REACTION

Cysteine and methionine are two amino acids that contain sulfur:

Cysteine	Methionine

Methionine is an essential amino acid, but *part* of the methionine requirement can be satisfied by cysteine. The sulfhydryl group ($-SH$) in cysteine reacts with sodium nitroprusside to give a red color:

Only those amino acids containing the $-SH$ group give the red color with nitroprusside.

4. REACTIONS WITH SALTS OF HEAVY METALS

Many proteins give *precipitates* with salts of heavy metals such as mercury (Hg^{2+}), lead (Pb^{2+}), silver (Ag^+), and copper (Cu^{2+}). This reaction is one of the processes that occurs in cases of acute toxicity of these metals when the salts are taken orally. In this case the metals react with the protein lining of the mouth and stomach and essentially destroy their normal functioning.

5. BIURET TEST

In basic solution copper ions react with many dissolved proteins to yield a pink to violet color that is the result of the formation of a copper complex. The biuret reaction is characteristic of molecules containing *two* ($-\overset{\parallel}{\underset{O}{C}}-NH-H$) groups joined through a single carbon or nitrogen atom or joined directly. Compounds in which the $-\overset{\parallel}{\underset{O}{C}}-NH-$ group is replaced by the $-\overset{\parallel}{\underset{S}{C}}-NH_2$,

$-\overset{\parallel}{\underset{NH}{C}}-NH_2$, or $-CH_2NH_2$ groups also gives this test. The test is named for the compound biuret,

$H_2N-\overset{\parallel}{\underset{O}{C}}-\overset{|}{\underset{H}{N}}-\overset{\parallel}{\underset{O}{C}}-NH_2$ which gives a positive result for the test. It is a characteristic test for proteins when they meet the structural requirements of the test.

6. MILLON'S TEST Proteins containing hydroxyphenyl groups $(-\!\!\bigcirc\!\!-OH)$ react with hot mercuric nitrate solutions to produce a red solution or, if precipitated, a red precipitate. This is essentially a test for the presence of *tyrosine* in the protein:

$$HO\!-\!\bigcirc\!-CH_2\!-\!\underset{NH_2}{CH}\!-\!\overset{O}{\underset{OH}{C}}$$

Tyrosine

7. DENATURATION AND COAGULATION OF PROTEINS The biologic functioning of a protein is generally dependent upon three types of structural features. The first is the sequence of amino acids in the chain; the second is the way parts of this amino acid chain interact with other parts via hydrogen bonding to form helices or sheets; and the third structural feature is the way the helices or sheets are folded to form larger structures. A protein can be rendered ineffective for the purpose for which it was synthesized by the organism if any structural feature is disrupted. Once this occurs the protein is described as **denatured.** The denaturation process may be reversible, depending upon how extensive it is. Proteins can be denatured by heat, light, strong acids, or alkalies, and by many other chemical and physical processes. In many cases, a denatured protein is reasonably soluble in water because of the excess of either positively or negatively charged groups on it. The relative numbers of charges change with pH. At a certain pH the numbers of positive and negative charges on the protein molecule are equal; this pH is called the **isoelectric point** and the protein generally has a minimum solubility at this point.

PROCEDURE

I. TEST FOR STRUCTURAL FEATURES Each of these tests is to be carried out on fresh solutions in clean test tubes.

Into a clean test tube place 2 mL to 3 mL of glycine solution; into another clean test tube put 2 mL to 3 mL of cysteine solution; into a third clean test tube place 1 mL to 2 mL of egg albumin solution; and into a fourth clean test tube place 1 mL to 2 mL of gelatin solution. Be sure to label the four test tubes (Fig. 37-1).

Figure 37-1. test tube arrangement for structural tests. A. Glycine solution. B. Cysteine solution. C. Egg albumin solution D. Gelatin solution.

Now add one reagent at a time (as specified below) to each of the test tubes. Swirl the contents gently to mix them and, when called for, heat them in a hot water bath (Fig. 37-2); then *record your observations on the data sheet*. When you have recorded your observations, clean the test tubes, add fresh portions of test solutions, and go on to the next test.

NINHYDRIN TEST: For this test add about 0.5 mL of the ninhydrin reagent solution to each of the solutions. Heat the solution gently to boiling and let it cool.

XANTHOPROTEIC TEST: Add 1 mL of concentrated nitric acid to your test solutions. Warm gently in the hot water bath to observe the complete sequence of reactions.

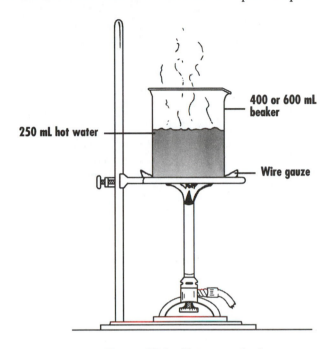

Figure 37-2. Hot water bath.

NITROPRUSSIDE REACTION: Add 5 drops of nitroprusside reagent solution and 2 mL of concentrated ammonium hydroxide solution to your test solutions.

MERCURIC REACTION: Add *dropwise* a few drops of mercuric test solution to your test solutions

BIURET TEST: Add 2 mL to 3 mL of concentrated sodium hydroxide to your test solution and mix thoroughly, then add dropwise about 1 mL of 0.5% copper solution with continuous mixing until a permanent color is obtained in the solution. A copper hydroxide precipitate will often form here. This is not to be confused with the pink to violet color that is positive for the Biuret test.

MILLON'S TEST: Add 3 drops of Millon's reagent and warm the test tube to the boiling point. Some proteins form an initial white precipitate which becomes red on heating, while others yield a red solution when treated similarly. Both are positive tests.

DENATURATION TESTS:
  With concentrated hydrochloric acid. Add a few drops of concentrated hydrochloric acid, mix thoroughly, and warm in the water bath.
  With concentrated sodium hydroxide. Add a few drops of concentrated sodium hydroxide, mix thoroughly, and warm in the water bath.

II. STRUCTURAL GROUPS FOUND TO BE PRESENT IN EGG ALBUMIN   *On the basis of your test results,* list the structural features which you found to be present in the protein of egg albumin.

III. STRUCTURAL GROUPS FOUND TO BE PRESENT IN GELATIN   *On the basis of your test results,* list the structural features which you found to be present in the protein of gelatin.

PRE-LAB QUESTIONS

1. Give the names for 3 chemicals used in this experiment which are especially harmful to the eyes.

    _____

2. List 3 body structures which are made up mostly of proteins.

    _____

3. What does a positive nitroprusside test indicate about a substance?

    _____

4. What are the two groups that all amino acids possess? These are the molecular features that most of the tests used in this experiment identify?

    _____

    _____

Experiment 37
Data Sheet

Name _____

Date _____

CHEMICAL PROPERTIES OF AMINO ACIDS AND PROTEINS

I. TEST FOR STRUCTURAL FEATURES—OBSERVATIONS:

	Glycine	Cysteine	Egg albumin	Gelatin
Ninhydrin reaction				
Xanthoproteic reaction				
Nitroprusside reaction				
Reaction with mercuric chloride solution				
Biuret test				
Millon's test				
Effect of concentrated hydrochloric acid				
Effect of concentrated sodium hydroxide solution				

II. STRUCTURAL GROUPS FOUND TO BE PRESENT IN EGG ALBUMIN:

III. STRUCTURAL GROUPS FOUND TO BE PRESENT IN GELATIN:

POST-LAB QUESTIONS

1. Most amino acids will give a positive ninhydrin test but most proteins will not. Explain why this is true.

   _____

2. What amino acid (as a constituent in a protein) in the skin reacts with nitric acid give a yellow color in the xanthoproteic test?

   _____

3. Explain what is meant by the term "isoelectric point" in reference to proteins.

   _____

38 The Isolation and Indentification of a Protein

OBJECTIVE

To isolate the protein casein from milk.
To examine some of the characteristic chemical properties of proteins.

APPARATUS AND CHEMICALS

250-mL conical flask
600-mL beaker
100-mL beaker
vacuum filtration apparatus
watch glass
test tubes
thermometer (110°C)
dropper
100-mL graduated cylinder
iron ring
ring stand
cheese cloth
acetic acid (glacial)

ether/ethanol mixture (1:1)
ethanol (95%)
3% solution of glycine in water
milk
wool
silk
nylon
$Pb(NO_3)_2$ solution (1%)
$Hg(NO_3)_2$ solution (1%)
$NaNO_3$ solution (1%)
NaOH solution (10%)
$CuSO_4$ solution (0.5%)
HNO_3, concentrated

SAFETY CONSIDERATIONS

 Wear eye protection to protect against possible shattering of dropped glassware and spillage of corrosive nitric acid and sodium hydroxide.

 Keep your hair and clothing away from open burner flame.

 Ether is very flammable. Use the ether in a hood, and keep all flames away. Discard as directed by your instructor.

 Nitric acid, HNO_3, and sodium hydroxide, NaOH, are corrosive Mercuric nitrate, $Hg((NO_3)_2$, and lead nitrate, $Pb(NO_3)_2$, are highly toxic metabolically. If any of these chemicals get on your skin, wash immediately under running water for several minutes.

FACTS TO KNOW

Proteins are large molecules found in the cells of living organisms and in biological fluids such as blood plasma. They contain the elements carbon, hydrogen, oxygen, nitrogen, and quite frequently sulfur. These huge molecules, the molecular weights of which range from 5000 to several million, are composed of approximately 20 different types of subunits known as amino acids, in the same way a freight train can be made up of as many as 20 or more different kinds of cars.

$$CH_3 - \underset{\underset{H}{|}}{\overset{\overset{NH_2}{|}}{C}} - \overset{\overset{O}{\|}}{C} - OH$$

The amino acid alanine

| Glycine | Alanine | Leucine | Isoleucine | Repeating Unit |

Pictoral representation of a protein

The properties and functions of proteins are exceedingly diverse. Some are relatively inert fibers, such as the keratins of wool and hair, or the collagens of tendon and connective tissue, which play an important structural role in animals. Others, such as those found in egg white and blood plasma, are globular and water soluble. One group of proteins, the enzymes, are the essential catalysts of biological systems. These enable reactions to proceed at a very rapid rate under very moderate conditions.

The chemical properties of the proteins allow us to identify and isolate them. The peptide bond is formed when an acid group undergoes a reaction with an amino group:

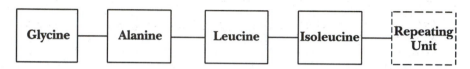

Peptide bond

Since each amino acid contains at least one acidic group and one amine or basic group, the protein chain may become very long, which of course gives rise to the high molecular weight chain. Some of the amino acids (e.g., lysine) contain more than one amine group while others (e.g., aspartic acid) contain more than one acidic group. These groups are not used in the propagation of the chain, but they do perform important functions. In fact, since there are acid and base groups in amino acids, the specific molecular structures are dependent on the basicity or acidity of the solution. When a large molecule has a net positive or negative charge, its solubility in water is enhanced. Conversely, when there are no charges or when the plus and minus charges are equal (isoelectric point), the protein is least soluble in water. This characteristic will be used in this experiment.

Lysine and aspartic acid in protein chain (at isoelectric point)

The principal protein found in milk is casein, which may be separated from the milk by acidification of a milk solution. This method causes the precipitation of the casein and some fat. The fat may be removed by washing the precipitate with alcohol because the fat is soluble in alcohol while the protein is not. The casein should give representative positive tests for protein.

PROCEDURE

Numbers in parentheses refer to entry numbers on the data sheet.

I. PREPARATION OF CASEIN   Place 100 g of milk in a 250-mL conical flask and heat in a water bath. When the temperature of the milk has reached 40°C, add dropwise, with stirring, about 20 drops of glacial acetic acid until all of the casein flocculates. This may be tested by observing whether there is any more precipitation upon addition of acid.

Filter the flocculate through cheese cloth and gently squeeze most of the water from the solid.

Place the casein-fat mixture in a 100-mL beaker containing 50 mL of 95% ethanol. After stirring, pour the liquid from the solid. Under a hood, the procedure is then repeated using 50 mL of a 1:1 ether-ethanol mixture. Stir for at least 5 minutes. Decant (pour off the liquid) into an appropriately labelled waste container in a hood. Add another 50 mL of 1:1 ether-ethanol mixture and stir for at least 5 minutes. Decant the liquid into the waste container in a hood. Transfer the remaining solid and liquid to a Büchner funnel equipped for vacuum filtration. The material adhering to the side of the beaker may be washed into the funnel with additional quantities of the 1:1 mixture of ether-ethanol.

After allowing the vacuum to remove most of the liquid, spread the casein on a watch glass. Chop up the casein with a spatula. Weigh the dried casein (1) and calculate the percent casein in the milk (2).

$$\% \text{ casein} = \text{Weight casein}/100 \text{ g milk} \times 100\%$$

II. IDENTIFYING CHEMICAL PROPERTIES

A. Heavy metals such as lead or mercury can cause serious damage to proteins. In the presence of these metals, the proteins precipitate (that is, the proteins become solid and settle out of solution).

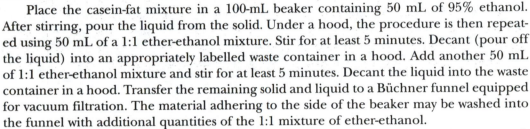

Test the action of the metals Pb^{2+}, Hg^{2+}, and Na^+ on milk. Arrange three test tubes in a test tube rack. To one test tube, add 10 drops of a 1% solution of Pb^{2+}; to another test tube, add 10 drops of Hg^{2+} solution; and, to the third test tube, add 10 drops of Na^+ solution. To each test tube, add 2 mL of milk. Shake each solution. Note any precipitate on the side of a test tube. It may take several minutes for any reaction to occur. Record the results for each metal (3, 4, and 5).

B. Another test that is frequently used for proteins is the biuret test. Add 1 mL of 10% sodium hydroxide solution to 1 mL of the milk. Then add slowly a very dilute (0.5%) copper sulfate solution until a purplish-violet color is produced (6). Repeat, except substitute 1 mL of a 3% glycine solution for the milk (7). The characteristic purple-blue or violet color of the intact protein is due to the presence of the peptide linkage. This test is commonly used to detect the presence of peptide linkages.

C. The xanthoproteic reaction confirms the presence of a special kind of amino acid. These, the aromatic amino acids, react with concentrated nitric acid to form intensely yellow compounds. Place a drop of concentrated nitric acid on the following substances and observe the reaction:

    (a) your dried casein (8)
    (b) wool fabric (9)
    (c) silk fabric (10)
    (d) nylon fabric (11)
    (e) cotton fabric (12)

Turn in the remainder of your product in a labelled bottle and indicate the percent casein in your milk sample.

PRE-LAB QUESTIONS

1. What characteristic does ether have that makes it dangerous in lab? What precaution should be taken in using ether?

    _____

2. List 5 elements found in protein molecules.

    _____

3. What is the most obvious difference between protein molecules and the common molecules encountered in other labs, such as water, ammonia, sugar, and alcohol?

    _____

    _____

4. What is the action of heavy metal ions on proteins?

    _____

    _____

Experiment 38
Data Sheet

Name _____

Date _____

THE ISOLATION AND IDENTIFICATION OF A PROTEIN

I. PREPARATION OF CASEIN

(1) Weight of dried casein _____ g

(2) Percent casein in sample _____ %

II. IDENTIFYING CHEMICAL PROPERTIES

A. Action of heavy metals

(3) Addition of Pb^{2+} to milk _____

(4) Addition of Hg^{2+} to milk _____

(5) Addition of Na^+ to milk _____

B. Biuret test

(6) Color observed on addition of copper
     sulfate to basic solution of milk _____

(7) Color observed on addition of copper
     sulfate to basic solution of glycine _____

C. Xanthoproteic reaction—addition of nitric acid to

(8) Casein _____

(9) Wool _____

(10) Silk _____

(11) Nylon _____

(12) Cotton _____

POST-LAB QUESTIONS

1. Name two types of functions carried out by proteins in the human body.

   _____

2. Acetic acid is added to milk to precipitate the protein casein. Exactly what does the acetic acid do chemically to cause the protein to precipitate?

   _____

3. What is the reason for adding ethanol and ethanol-ether solutions to the precipitate?

   _____

4. Would it be easier to recover casein from skim milk or from whole milk? Explain.

   _____

EXPERIMENT 39

Titration of Magnesium Salts in Hard Water

OBJECTIVE

To demonstrate how magnesium salts in water may be determined by titration with the chelating agent, EDTA (ethylenediaminetetraacetic acid). Similar techniques may be used for the determination of the concentration of calcium salts in water.

APPARATUS AND CHEMICALS

25-mL pipet
rubber bulb for pipet
50-mL buret
.01 M Na₂EDTA (disodium ethylenediaminetetraacetic acid)

.01 M MgCl₂
pH 10 buffer
Eriochrome Black T indicator solution
100-mL beaker or 250-mL conical flask

SAFETY CONSIDERATIONS

Wear eye protection at all times.

FACTS TO KNOW

Magnesium ion is one of many ions whose concentration can be determined by the use of a complexometric titration. In this determination the Eriochrome Black T solution is added to the buffered solution of magnesium salt. This results in the formation of a *red* complex between the magnesium and the Eriochrome Black T:

$$Mg^{2+} + \quad \rightleftharpoons \quad + 2H^+$$

Magnesium ion   Eriochrome Black T   Magnesium Eriochrome Black T complex
                 (blue)   (red)

Magnesium ion forms a much more stable complex with EDTA

$$HOOCCH_2 \diagdown N - CH_2 CH_2 - N \diagup CH_2 COOH$$
$$HOOCCH_2 \diagup \qquad \diagdown CH_2 COOH$$

EDTA

than it does with the dye Eriochrome Black T. As a result, the addition of EDTA solution to a solution which contains both magnesium ion and the magnesium Eriochrome Black T complex will result first in a complex between the free magnesium ion and the EDTA. After the free magnesium ion is complexed, the EDTA extracts the magnesium present in the dye complex. Thus at the endpoint the EDTA has tied up all the magnesium, and the solution no longer has the red color characteristic of the magnesium-dye complex, but rather the blue color due to the free dye itself. The reaction with EDTA may be written as

$$Mg^{2+} + (Mg^{2+} - Dye) + EDTA \rightarrow MgEDTA + Dye$$
$$\text{(free)} \qquad \text{red} \qquad\qquad\qquad\qquad \text{blue}$$

The essential feature of this reaction is that one EDTA molecule is used for *each* Mg^{2+} ion present in the original solution.

Hard water contains dissolved salts which react with soap to give a precipitate. Hard water can be formed when ground waters percolate through limestone ($CaCO_3$) or dolomite ($CaCO_3 \cdot MgCO_3$) or when water dissolves calcium sulfate or magnesium sulfate. Water containing dissolved iron salts is also hard water. Hard water causes problems in any process in which the dissolved salts form a precipitate (including laundering) and in steam generation in boilers.

PROCEDURE

Obtain a clean 50-mL buret and rinse it with small amounts of the Na_2EDTA solution. Then fill the buret with the Na_2EDTA solution and bring the level in the buret to the zero mL mark (1). Then use a 25-mL pipet with rubber bulb to pipet 25 mL of the magnesium chloride solution (4) into a 100-mL beaker or 250-mL conical flask. Dilute to 50 mL with deionized or distilled water, and add 10 mL of the buffer solution to bring the solution to a pH of 10. Stir or swirl this solution gently while adding the Eriochrome Black T solution to get an easily visible color (10 to 15 drops should be sufficient).

Titrate this solution carefully with the Na_2EDTA solution until the color of the solution changes from red to purple to blue. This must be done by adding small amounts of the Na_2EDTA solution to the magnesium chloride solution and then swirling the mixture until the solution changes to a uniform color throughout. Near the endpoint the Na_2EDTA solution should be added dropwise. As soon as the color of the solution has changed to a uniform blue, stop the titration and record the volume of Na_2EDTA solution used (2).

Repeat the titration after you have refilled the buret with Na_2EDTA solution and brought the initial volume to the zero mL line.

EXAMPLE CALCULATION

Suppose that 26.15 mL of Na_2EDTA is required to obtain the necessary color change and that the concentration of Na_2EDTA is .0100 M. Find the concentration of Mg^{2+} in the original sample of 25 mL of magnesium chloride solution taken for analysis.

Since one mole of Na_2EDTA is required to react with each mole of Mg^{2+}, we can write

$$\text{Concentration of } Mg^{2+} = 0.0100 \text{ M} \times \frac{26.15 \text{ mL}}{25.00 \text{ mL}} = .0104 \text{ M}$$

PRE-LAB QUESTIONS

1. How do metals such as calcium and magnesium get into our water supply?

    _____

    _____

    _____

2. Explain how Eriochrome Black T works as indicator in this titration.

    _____

    _____

    _____

3. What color is the free or uncomplexed Eriochrome Black T?

    _____

4. When taking a bath in hard water, what is the residue (scum) that may be found on the sides of the bathtub?

    _____

    _____

    _____

Experiment 39
Data Sheet

Name _____

Date _____

TITRATION OF MAGNESIUM SALTS IN HARD WATER

Concentration of the standard Na_2EDTA solution _____ molar.

	Titration No. 1	**Titration No. 2**
Na_2EDTA solution buret:		
(1) Initial reading	_____ mL	_____ mL
(2) Final reading	_____ mL	_____ mL
(3) Volume used	_____ mL	_____ mL
(4) $MgCl_2$ solution: volume taken	_____ mL	_____ mL
(5) Concentration of Mg^{2+} in unknown	_____ M	_____ M
(6) Average concentration Mg^{2+} in solution	_____ M	

Do your calculations below, showing clearly the reasoning used.

POST-LAB QUESTIONS

1. EDTA is sometimes used therapeutically to remove high levels of lead in the blood. Explain how this might work.

    _____

    _____

    _____

2. Since detergents sometimes react with the metals in hard water to produce a precipitate, explain what the effect of adding some EDTA to the detergent formulation.

    _____

    _____

    _____

3. Suppose that 32.7 mL of 0.02 M Na_2EDTA was required to completely react with 42.0 ml of a magnesium chloride solution. What would be the concentration of magnesium? Show calculations.

4. Even though you are determining the amount of magnesium ion in the water, you add an indicator that actually *contains* magnesium. Explain why this is necessary for the analysis.

    _____

    _____

    _____

40 Chemical Oxygen Demand of Polluted Water

OBJECTIVE

To determine the chemical oxygen demand (COD) of water contaminated with organic waste materials.

APPARATUS AND CHEMICALS

250-mL boiling flasks with ground glass necks (2)
water-cooled condensers with ground glass joints (2)
hot plates or Bunsen burners (2)
20-mL pipet
10-mL pipet
rubber bulb for pipet
50-mL graduated cylinder
glass beads
50-mL buret
0.0417 M potassium dichromate ($K_2Cr_2O_7$) solution

H_2SO_4, concentrated, reagent grade
H_2SO_4, concentrated, reagent grade, containing 22 g of Ag_2SO_4 (silver sulfate) per 9 pounds of H_2SO_4
0.10 M ferrous ammonium sulfate hexahydrate ($Fe(NH_4)_2(SO_4)_2 \cdot 6H_2O$)
ferroin indicator solution
$HgSO_4$ (mercuric sulfate), analytical grade crystals
glucose solution
polluted water

SAFETY CONSIDERATIONS

 When using a burner in the laboratory, be absolutely sure that loose hair and clothing do not come close to the flame.

 Be sure to wear eye protection. Care should be exercised when using concentrated H_2SO_4, and during the refluxing operations. Should skin or eye contact occur rinse with lots of water for at least 15 minutes (eyes).

 Be cautious when using the burner and do not touch the glassware or support mechanisms while they are hot.

 Do not spill concentrated H_2SO_4 on clothing. Gloves and a laboratory coat will provide additional safety.

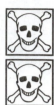

Potassium dichromate and mercury(II) sulfate are poisonous.

Be sure to use a rubber bulb in the pipet transfer of polluted water or unknown sample. Your instructor will demonstrate the technique.

FACTS TO KNOW

Polluted water often contains organic matter which can come from a number of sources—improperly treated sewage, discharge from industrial plants, leakage from vessels or pipelines transporting chemicals or petroleum products, and other sources. Such pollution can alter the smell and taste of the water and thus lower its quality, especially for drinking purposes. Harmful effects to fish and other aquatic life can also result from the presence of excess organic matter.

Some pollutants are toxic substances that poison living organisms. Other pollutants are problems for a different reason. Bacteria in natural environments can use these pollutants as food, and as the bacteria consume them, the pollutants are broken down to harmless products. This is the way in which natural waters cleanse themselves, and is the manner in which the world was meant to work. The problem comes when the amount of the pollutants is so large that it puts a stress on the natural system's ability to deal with them. As the bacteria consume the pollutants, they also utilize the dissolved oxygen in the water as a part of the process. When large amounts of pollutants are present, the bacteria use so much of the dissolved oxygen in the water that other aquatic life cannot survive on the remaining oxygen, and fish kills as well as harm to other aquatic life results from this oxygen depletion.

This experiment illustrates how one measures the amount of dissolved oxygen that the organic waste present in the water needs in order to be degraded to harmless products. Rather than using bacteria for the degradation, however, the procedure uses a chemical oxidizing agent to accomplish the same reaction—hence the name "chemical oxygen demand" or COD.

This is a standard method for determining the quantitative extent of pollution by organic materials and involves oxidizing the organic substances with a strong chemical oxidizing agent. In this experiment a potassium dichromate-sulfuric acid mixture is used. Silver sulfate is added to catalyze the oxidation of these substances. The excess dichromate is titrated with ferrous ion according to the equation

$$6\,Fe^{2+} + Cr_2O_7^{2-} + 14\,H^+ \longrightarrow 6\,Fe^{3+} + 2\,Cr^{3+} + 7\,H_2O$$

This method, with the concentrations used, is most accurate when COD values are 50 ppm or more. Values in the vicinity of 10 ppm or less are best determined with dichromate and ferrous ammonium sulfate concentrations of 0.00417 M and 0.01 M, respectively (one tenth as concentrated as in the procedure described in this experiment).

PROCEDURE

Label two reflux flasks #1 and #2. Place about 0.4 g of $HgSO_4$ into flask #1. Add 20.0 mL of polluted water (or unknown provided by the instructor) to the flask by pipet (Fig. 40-1) and mix by swirling. Then add 10.0 mL of standard $K_2Cr_2O_7$ solution by pipet. Carefully measure out 30 mL of concentrated H_2SO_4, containing Ag_2SO_4, in a graduated cylinder and add to reflux flask #1. Add several glass beads to prevent bumping. (*CAUTION!!* The reflux mixture *MUST* be thoroughly mixed before heat is applied. If the mixing is not thorough, the extremely corrosive mixture may be blown out of the condenser when heat is applied.)

Make up a "blank," or reference solution, in flask #2 by adding all the ingredients above EXCEPT that instead of polluted water (or unknown) add 20.0 mL of distilled water. This blank solution serves as a comparison to avoid accidental errors.

Attach the reflux condensers to flasks #1 and #2 and start the water flow, as indicated in Figure 40-2. Apply sufficient heat to boil the contents and continue heating for 1 hour. Cool and wash down the inside of each condenser by pouring a few milliliters of distilled water into the top of the condenser. Dilute the mixture in each flask to about 140 mL with distilled water and cool to room temperature. Titrate the unreacted dichromate solutions with standard ferrous ammonium sulfate, adding 2 or 3 drops of ferroin indicator solution to each titrating flask *before beginning the titration.* (See Fig. 40-3 for the titration set-up.) When sufficient ferrous ammonium sulfate has been added, the color of the solution will suddenly change from blue-green to reddish brown. This color change should be taken as the endpoint even though the blue-green may reappear within minutes.

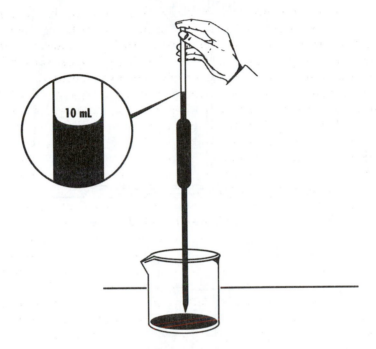

Figure 40-1. Meniscus alignment in use of pipet.

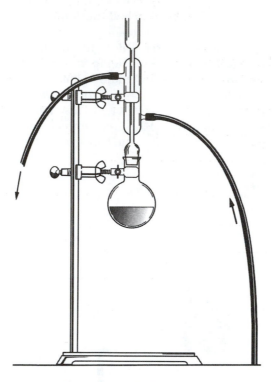

Figure 40-2. Reflux set-up.

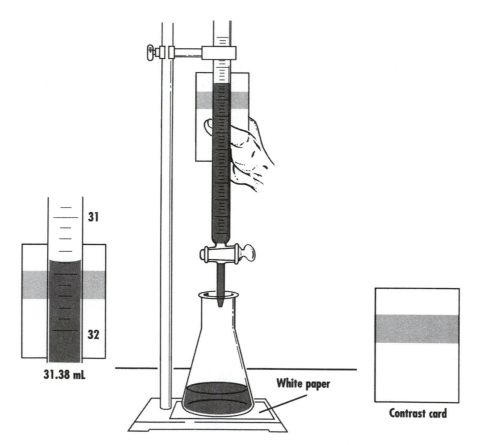

31

32

31.38 mL

White paper

Contrast card

Figure 40-3. Equipment arranged for a titration; use of contrast card in reading buret.

PRE-LAB QUESTIONS

1. The chemicals used in this experiment are dangerous and proper precautions should be observed. In the event that you do get some of the acid in your eyes, what should be your first aid procedure?

    _____

    _____

    _____

2. What is the primary problem with nontoxic organic waste materials in water?

    _____

    _____

    _____

3. What is the purpose of running a blank in this experiment?

    _____

    _____

    _____

4. Why is it necessary to insure that the sulfuric acid-silver sulfate solution is thoroughly mixed with waste water before you begin heating it?

    _____

    _____

    _____

Experiment 40
Data Sheet

Name _____

Date _____

CHEMICAL OXYGEN DEMAND OF POLLUTED WATER

(1) mL $Fe(NH_4)_2(SO_4)_2$ used for blank

_____ mL

(2) mL $Fe(NH_4)_2(SO_4)_2$ used for sample mL

_____ mL

CALCULATIONS

$$\text{ppm COD} = \frac{(a - b)c \times 8000}{\text{mL sample}}$$

where COD = chemical oxygen demand in parts per million

a = mL $Fe(NH_4)_2(SO_4)_2 \cdot 6H_2O$ for blank

b = mL $Fe(NH_4)_2(SO_4)_2 \cdot 6H_2O$ for sample

c = molarity of $Fe(NH_4)_2(SO_4)_2$

(3) ppm COD for polluted water

_____

POST-LAB QUESTIONS

1. Silver sulfate is added to catalyze the oxidation of the organic material. What is a catalyst?

    _____

    _____

    _____

2. What is the purpose of the condenser in the reflux operation?

    _____

    _____

    _____

3. Why is it necessary to have an excess of sodium dichromate in the reaction vessel?

    _____

    _____

    _____

4. COD is reported in parts per million (ppm). If you got a COD of 100 ppm in your experiment, this means 1,000,000 pounds of the water mixture would have 100 pounds of organic COD. Many municipalities charge industry on the basis of the number of pounds of organic waste discharged to the sewer. Suppose company, ABC discharged 1,000 gallons of water per day to the sewer. The COD of the discharge was found to be 150 ppm. Assuming that one gallon of water weighs 8.34 pounds, calculate the number of pounds of organic material received by the sewer in a day. Show Calculations.

41 Preparation of a Face Cream

OBJECTIVE

To become familiar with the composition of a face cream.
To deduce the purpose of some components in face cream.

APPARATUS AND CHEMICALS

150-mL beaker
400-mL beaker
stirring rod
water bath
burner
ring

110°C thermometer
ring stand
stearic acid
lanolin
mineral oil
triethanolamine

SAFETY CONSIDERATIONS

 Some components of the face creams are "greasy" or slippery and might make handling the glassware difficult. Take care not to allow the beaker to slip out of your grasp or it may shatter.

 When using a burner or any open flame in the laboratory, be absolutely sure that loose hair and clothing do not come close to the flame. It is easy to forget this simple rule when concentrating on measurements.

 Boiling water can splash into your eyes and cause permanent damage. Wear your safety glasses.

 Be sure the gas is turned off when you are through with it since in a closed laboratory a small gas leak can create an explosive mixture.

FACTS TO KNOW

Face creams are healthful only if they aid the skin in carrying out its normal functions—that is, the elimination of waste matter and the cooling of the body by radiation, thus aiding in the maintenance of the normal body temperature—and if they prevent harshness and a dry condition of the skin. A cream which clogs the pores of the skin with heavy, insoluble, inert material is detrimental to health.

Face creams contain a variety of ingredients. A satisfactory face cream can be prepared from stearic acid, lanolin, mineral oil, triethanolamine, and water. It is the purpose of this experiment to prepare a face cream from these ingredients, and by leaving an ingredient out in subsequent preparations to deduce the purpose of that particular ingredient. Face creams contain ingredients for adding body, improving texture, emulsifying the oil and water components, raising the melting point, improving the spreadability, improving the odor, softening the skin, and providing various medicinal properties.

PROCEDURE

Prepare 4 cold creams as follows. Use the ingredients and amounts shown in Table 41-1. Label each mixture.

TABLE 41-1

	MIXTURE 1	MIXTURE 2	MIXTURE 3	MIXTURE 4
Stearic acid	5 g	5 g	5 g	
Lanolin	3.5 g	3.5 g	3.5 g	3.5 g
Mineral oil	5 g	5 g	—	5 g
Triethanolamine	1 mL	—	1 mL	1 mL
Water	24 mL	24 mL	24 mL	24 mL

Place the stearic acid, lanolin, and mineral oil in a 150-mL beaker and heat on a water bath until all the ingredients have melted. (Cosmetic ingredients should not be melted over a direct flame because if they are heated much above the boiling point of water, they may scorch or decompose.)

Place the triethanolamine in 24 mL of water in a 400-mL beaker and heat on the water bath.

After the solutions have reached 80°C or 90°C (after about 4 minutes in boiling water), **slowly** pour the **oil into the water** a little at a time, stirring constantly. If you pour too fast or if you do not stir, your emulsion will be lumpy. Continue stirring until you have a smooth, uniform paste. The length of time it takes to achieve this with mixture 1 indicates how long you should stir the other mixtures.

Compare the properties of each cream and record your observations on the data sheet. Note which ingredient is missing in each preparation. Ascertain the function of the missing ingredient based on the difference in properties between a normal cold cream and the other preparations.

Obtain an unknown. Compare the properties of your unknown with the properties of your preparations and determine which ingredient has been omitted or reduced in amount.

PRE-LAB QUESTIONS

1. What precautions should be taken in using the Bunsen burner and water bath in this experiment?

    _____

2. List 2 functions of skin.

    _____

3. In this experiment, the functions of different components of face powder are investigated. What is done to assess the function of each ingredient?

    _____

    _____

    _____

Experiment 41
Data Sheet

Name _____

Date _____

PREPARATION OF A FACE CREAM

Function of the Ingredients:

Stearic Acid

Mineral Oil

Triethanolamine

Ingredient missing from unknown and reason for your choice:

POST-LAB QUESTIONS

1. Cosmetic ingredients should not be heated over an open flame. Why?

    _____

2. What characteristic was missing from your face cream when mineral oil was omitted?

    _____

3. Lanolin is called an "emollient." From your knowledge of the type of products in which lanolin is found, what do you believe is the function of an emollient?

    _____

    _____

    _____

4. Almost all commercial face creams contain one rather obvious component that was not added to any of the test creams. What might this be? (Hint—what is a very noticeable feature of the tri-ethanolamine?)

    _____

    _____

    _____

42 Analysis of a Face Powder

OBJECTIVE

To determine the substances that compose a typical face powder.

APPARATUS AND CHEMICALS

10-mL or 25-mL graduated cylinder
funnel
test tubes (6)
U-shaped glass tubing for
   carbonate test
evaporating dish
burner
mortar and pestle
blow pipe
face powder sample
phenolphthalein
ethanol, denatured
blue litmus paper
red litmus paper
calcium hydroxide,
   saturated solution
filter paper to fit funnel
medicine droppers (2)

HCl, concentrated
HCl, dilute (3 M – 6 M)
ammonium hydroxide (6 M)
   (dropper bottle)
ammonium oxalate solution
acetic acid (6 M)
ammonium chloride solution
ammonium carbonate solution
disodium phosphate solution
sodium sulfide solution
charcoal block
sodium carbonate, solid
cobalt nitrate solution
potassium hydrogen sulfate, solid
nitric acid, concentrated
stannous chloride solution
   (freshly prepared)
iodine solution

SAFETY CONSIDERATIONS

The alcohol used in this experiment has been denatured in that a poison has been added so that it can not be used for beverage alcohol.

When using the assembly in Figure 45-1, be sure that the system is open so that gas can escape, because a relatively large amount of gas will be produced by a small amount of carbonate and hydrochloric acid.

The dilute hydrochloric acid can damage your clothing and irritate your skin.

Concentrated hydrochloric acid and concentrated nitric acid are dangerous to your skin, nose, and eyes. Be sure to use the hood for the transfer of these chemicals.

Treat solutions of metal ions, such as zinc, as poisons. Clean up the area as directed and wash at the end of your work.

A solution of sodium sulfide is quite alkaline and is, as a result, a corrosive poison. Do not allow this solution to come into contact with an acid nor with your skin.

It is easy to burn yourself when learning techniques with a blow pipe. Be sure the charcoal is not burning and place it for storage under the directions of your instructor. Loose hair is especially dangerous here. Be sure the gas is turned off completely at the end of the work.

The stannous chloride solution is a strong acid solution.

Iodine solutions will burn your skin if not washed away.

FACTS TO KNOW

Face powders are used by both men and women. Women use them to achieve a desired color on the face, to reduce greasiness, to improve texture, and to mask blemishes. Men use face powder for some or all of the same reasons, and in addition, for protective and antiseptic purposes, such as after shaving.

Face powders can be harmful if they contain ingredients that will clog pores in the skin and prevent elimination of waste matter and cooling of the body by radiation. Such powders may contain starch, powdered talc, carmine, and perfume; some contain zinc oxide. Face powders are perhaps more harmful if they contain a substance that will be absorbed through the skin and cause a toxic reaction. Lead and mercury compounds are particularly harmful when found in face powders and creams. Both lead and mercury compounds may be absorbed through the skin easily and cause poisoning. When basic lead carbonate ($2 PbCO_3 \cdot Pb(OH)_2$) is absorbed, it produces a type of lead poisoning. Mercury compounds are sometimes found in lotions and powders used to remove freckles. It should be understood, however, that poisonous ingredients will not be found in preparations prepared by reputable manufacturers.

Face powders contain a variety of inorganic solids for body and some for antiseptic action (boric acid, H_3BO_3, for example), a coloring material, and a perfume. The body is afforded by such compounds as magnesium carbonate ($MgCO_3$), boric acid, starch ($(C_5H_{10}O_5)_n$), talc ($Mg_3Si_4O_{10}(OH)_2$), calcium carbonate ($CaCO_3$), and zinc oxide (ZnO).

You will make a test on either a commercial face powder or one prepared for this experiment. This experiment is not performed for the purpose of detecting harmful substances, but rather to determine the composition of a typical commercial face powder.

PROCEDURE

Numbers in parentheses refer to entry numbers on the data sheet.

Ration your sample of face powder so you will have enough for each test and a little left over for a check on results that are less certain.

TEST FOR ALKALI   To test for both free alkali and alkali formed by hydrolysis, place a small amount of your sample in 5 mL of water. Shake the tube vigorously. Add 3 drops of phenolphthalein. Note any color change (1). A pink color indicates the presence of alkali.

To test for free alkali, add a small amount of your sample to 5 mL of alcohol. Shake the tube vigorously and add 3 drops of phenolphthalein. Note any color change (2).

Does your sample contain free alkali (3)?
Does your sample contain alkali formed by hydrolysis (4)?
What danger can be caused by alkali on the skin (5)?

TEST FOR CARBONATE   Place a small amount of the sample in a dry test tube. Assemble the apparatus as shown in Figure 42-1. (The calcium hydroxide solution MUST BE CLEAR; you may have to filter or centrifuge it before using.)

Add dilute hydrochloric acid to the tube containing your sample and quickly replace the stopper. A precipitate (cloudiness) in the calcium hydroxide solution indicates the presence of carbonate. Is carbonate present in your sample (6)?

Supply the other product in the reaction (7):

$$CaOH_2 + CO_2 \rightarrow \underline{\hspace{4cm}} + H_2O$$

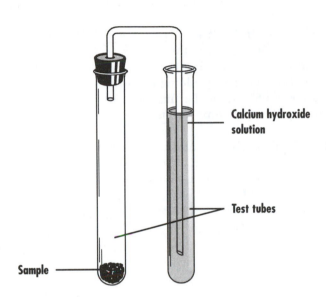

Figure 42-1. Assembly for testing for a carbonate by the evolution of carbon dioxide.

Calcium hydroxide solution

Test tubes

Sample

TEST FOR CALCIUM ION   Place a small amount of your sample in an evaporating dish and heat UNDER THE HOOD with 5 mL of concentrated hydrochloric acid. Dilute with 10 mL of distilled water and filter. Divide the filtrate into three equal portions. Use one portion to test for calcium, one for magnesium, and the third for zinc.

To test for calcium, transfer a portion of the solution just prepared to a small beaker and add dilute ammonium hydroxide drop by drop until the solution becomes slightly alkaline (until red litmus paper just turns blue). Heat almost to boiling and add 3 mL of ammonium oxalate solution. A fine, white crystalline precipitate, insoluble in acetic acid and soluble in dilute hydrochloric acid, indicates the presence of the calcium ion. Record the results (8).

Write the formula for the possible precipitate by filling in the equation (9):

$$CaCl_2 + (NH_4)_2C_2O_4 \rightarrow 2\ NH_4Cl + \underline{\hspace{4cm}}$$

TEST FOR MAGNESIUM ION   Remove the calcium ion, if present, by making a second portion of the solution alkaline. Add first 3 mL of ammonium chloride and then ammonium hydroxide, a few drops at a time, until litmus paper turns blue. Add 2 mL of ammonium carbonate solution and filter off the calcium carbonate precipitate.

To the clear filtrate (or to the original second portion if no calcium is present) add 2 mL of disodium phosphate solution. A fine, white crystalline precipitate that forms very slowly indicates the presence of the magnesium ion. Record the results (10).

TEST FOR ZINC ION Make the third portion of solution alkaline with dilute ammonium hydroxide and add 2 mL of sodium sulfide solution. A white precipitate indicates zinc. Record your results (11).

(This test fails when metallic ions which give dark-colored sulfides are present. If this is the case, mix some of the original sample of face powder with solid sodium carbonate by grinding the two together in a mortar. Place a small amount of this mixture in an indentation of a charcoal block, moisten with a drop of cobalt nitrate solution, and heat, using a blowpipe to direct the flame on the mixture. A green color indicates the presence of zinc. Ask the instructor to demonstrate this technique.)

TEST FOR BORIC ACID AND BORAX Mix 2 g of your sample with 2 g of potassium hydrogen sulfate. Grind well in a mortar, transfer the mixture to an evaporating dish, add 5 mL of ethanol, stir, and ignite carefully. A green flame indicates the presence of some borate or boric acid. What are your results (12)?

TEST FOR MERCURY ION *UNDER A HOOD,* place about 2 g of your sample in an evaporating dish, cover the sample with concentrated nitric acid, and heat for a few minutes. Let the solution cool. Add an equal volume of water and filter. Test the clear filtrate for mercury by adding a few drops of freshly prepared stannous chloride solution to a few milliliters of the filtrate. A white or gray precipitate indicates the presence of mercury. What are your results (13)?

TEST FOR STARCH Add a drop or two of iodine solution to a small portion of your solid sample. Let it set for a minute or so and observe any color change. If starch is present, the iodine will cause a dark blue or black coloration. Describe the effect of iodine on your sample (14).

Summarize your results by listing the substances that are *present* in your sample and those that are *absent* (15).

PRE-LAB QUESTIONS

1. What is meant by "denatured alcohol" and what precaution should be used in handling it in the lab?

    _____

2. If an acid or base solution used in this experiment is marked "dilute" does that mean it cannot be harmful to the skin or eyes?

    _____

3. How does one test for the presence of starch in the face powder? This is a widely used test, as starch is a component of many natural and synthetic products.

    _____

    _____

    _____

Experiment 42
Data Sheet

Name _____

Date _____

ANALYSIS OF A FACE POWDER

(1) _____

(2) _____

(3) _____

(4) _____

(5) _____

_____

(6) _____

_____

(7) _____

(8) _____

_____

(9) _____

(10) _____

_____

(11) _____

_____

(12) _____

_____

(13) _____

_____

(14) _____

(15) Substances found to be present:

Substances found to be absent:

POST-LAB QUESTIONS

1. What color will each indicator below exhibit in basic (alkaline) solution?

    (a) phenolphthalein _____

    (b) litmus paper _____

2. Name the purpose for each ingredient below in face powders.

    (a) starch

    (b) boric acid

    (c) zinc oxide

3. If phenolphthalein is added to a solution and the color turns pink or red, what substance is present?

    _____